THE
CHALLENGE TO AI

THE
CHALLENGE TO AI
*Consciousness and Ecological
General Intelligence*

STEPHEN E. ROBBINS, PhD

MERCURY LEARNING AND INFORMATION
Boston, Massachusetts

Publisher: David Pallai
MERCURY LEARNING AND INFORMATION
121 High Street, 3rd Floor
Boston, MA 02110
info@merclearning.com
www.merclearning.com
800-232-0223

S. E. Robbins. *The Challenge to AI: Consciousness and Ecological General Intelligence.*
ISBN: 978-1-50152-191-1

The publisher recognizes and respects all marks used by companies, manufacturers,
and developers as a means to distinguish their products. All brand names and product
names mentioned in this book are trademarks or service marks of their respective
companies. Any omission or misuse (of any kind) of service marks or trademarks, etc.
is not an attempt to infringe on the property of others.

Library of Congress Control Number: 2023950395

242526321   This book is printed on acid-free paper in the United States of America.

Our titles are available for adoption, license, or bulk purchase by institutions,
corporations, etc. For additional information, please contact the Customer Service
Dept. at 800-232-0223(toll free).

All of our titles are available in digital format at academiccourseware.com and other digital
vendors. The sole obligation of MERCURY LEARNING AND INFORMATION to the purchaser is to
replace the files, based on defective materials or faulty workmanship, but not based on the
operation or functionality of the product.

To Susan, the love and support of this work.

CONTENTS

PREFACE

In 1979, as the advance on artificial intelligence was really gaining force, there appeared *Gödel, Escher, Bach: an Eternal Golden Braid*, or "GEB" for short. This was Douglas Hofstadter's great contemplation on mind and computers with just a grain or two of Zen thrown in. Hofstadter has since become one of the great skeptics on whether AI can ever master language, and while it seems that his defensive position is being shrunk, so to speak, into an ever-decreasing perimeter, the present book will endeavor to show that there are deeper considerations that Hofstadter could have made, starting, in fact, with the nature of space and time.

There was a time when the title of the present book was craftily (it was thought) contemplated to be another "GEB," playing off Hofstadter's work, namely, "Gibson, Einstein, Bergson: Another Golden Braid," while also being a book on the nature of mind and the nature of AI. The name, Einstein, of course, easily evokes for everyone the subject of space and time. Einstein's contemporary, the French philosopher, Henri Bergson, however, is almost a forgotten name, even in philosophy, yet his reputation as a "philosopher of time" was so high in those days and his dispute with Einstein over the interpretation of Special Relativity and the famous twin-paradox so salient (the paradox wherein a rocket-traveling twin is held to age less than his brother who remains on the earth), that it is suspected this debate might have resulted in Einstein receiving the Nobel Prize not for relativity, but rather for the photoelectric effect.

This debate, in our opinion, was far from resolved. It goes to the essence of time, and time and consciousness are inextricably linked. So at the beginning of the AI era, it came as a shock to the author when as a graduate student, he discovered that the great theorist of perception, James J. Gibson, sided with Bergson and against the physicists. Gibson is the founder of "ecological psychology." He is the theorist of "affordances," these being the very essence of ecological intelligence, and also the proponent of "direct perception," wherein we *directly* see the coffee cup "out there" on the kitchen table, there being no image of the cup "in the brain," or in some "mental space." A remark Gibson made in a conference talk, to wit, "Physicists mislead us when they say 'there is no simultaneity'..." came after reading a paper by the author attempting to reconcile Bergson and Einstein. Gibson is right. We shall see that to understand consciousness, conscious perception, *sentience*— that simple question of how we see, thus experience the coffee cup "out there" on the kitchen table, spoon stirring, liquid surface swirling—we must discard physics concepts of space and time; we require a different framework, one that had been explicitly proposed by Bergson and that is required for understanding Gibson.

And so this book will dive into the source of the profound difference between man and machine, between AI language understanding and human language understanding, between AI "perception," and human perception, and into the basis for general intelligence. In this, we will see, finally, an answer as to why indeed "the biology is important," and why, although hitherto regarded as unrelated, even an unnecessary relation, consciousness is required for cognition. The "device" (thinking of how we characterize the body and brain) required to support all this is quite different from current conceptions. To achieve it, to *engineer* it—for it requires hardware, not software—will indeed be a challenge for AI.

S. E. Robbins
December 2023

CHAPTER 1

INTRODUCTION: AGI—THE GLEAM IN THE EYE OF AI

THE COMMONSENSE MOUSETRAP

Artificial general intelligence (AGI), is the gleam in the eye of the AI community. This gleam was there "at the beginning," the beginning being in the 1950s. However, in the author's experience (circa 1972) the computer metaphor of mind swept into a then rather sleepy cognitive psychology just barely emerging from the grips of Skinner's rats, a field that, given this injection of the computational framework, ultimately became *cognitive science*. This beginning was with the publication of Newell and Simon's *Human Problem Solving*, a work dedicated to showing that a fundamental algorithm, *means–ends analysis*, could be applied to tasks like theorem proving, cryptarithmetic problems (e.g., given $D = 5$, show DONALD + GERALD = ROBERT), even chess. It was dedicated to a human-simulation quest, that is, to showing that human solution processes for the problem closely emulated the steps taken by Newell and Simon's computer algorithm (which they termed the General Problem Solver [GPS]). This algorithmic scheme, it turned out, would eventually evolve, becoming the industry of "expert systems."

While the GPS tasks previously noted do not sound too much like commonsense or general intelligence, in 1971 there was an article by Freeman and Newell titled "Functional Reasoning in Design" [Freeman1971]. This explored an algorithm, very much like means–ends analysis, that could *design* objects, in their example, a knife. Here they envisioned a database listing "functional provisions" for objects, and "functional requirements." The object, "handle" would *provide* the function "holding," the object, "blade" would *provide* "cutting," and the blade would have a functional requirement as well, namely, "being held." By searching through the database and "matching" the requirements to the provisions, the program could *design* a knife—the handle providing holding for the blade, the blade providing "cutting." So, indeed, we would be right there in the Stone Age era with guys dressed in bearskins figuring out how to fashion knives—commonsense knowledge and reasoning to be sure, in fact, something even very "ecological." Ecological psychology—the school of the great theorist of perception, J. J. Gibson—would use the term *affordance* here, for surely we are perceiving that the handle "affords" holding and the blade affords cutting things, for example, deerskins.

Let us take this dimension of commonsense intelligence a bit further. Earlier, perhaps in the 1960s (the author no longer can locate the precise issue), in the widely read magazine *Readers Digest*, a little "creativity test" for engineers was featured. Suppose, the test proposed, you were given a set of objects: a small box (say, 9" × 9" × 9"), razor blade, toothpicks, rubber bands, string, paper clips, stapler, some sharpened pencils, a chunk (of course) of cheese from the author's home state of Wisconsin: your task—design a mousetrap. We will proceed with this task via some analogies. First, we will use a sharpened pencil to form a sort of crossbow. The pencil sticks through a hole made in the box wall, it is drawn back with rubber bands stapled to the side, then a toothpick is lodged in a groove to hold the pencil in place, and a string is tied to the toothpick and then to the cheese (Figure 1.1). Or perhaps we design a mouse "beheader" by lodging a razorblade into a pencil, lodging the sharp end of the pencil into a box corner, propping it up with a toothpick, and using rubber bands for pull-down force.

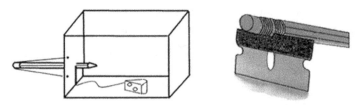

FIGURE 1.1. Left: A "crossbow" trap. Right: Part of a "beheader" trap.

It would seem that Freeman and Newell's database of functional provisions and functional requirements (presumably having entries for each of the objects in the potential mouse trap set) is in trouble. Did we really remember to list "piercing" as a functional provision of the pencil? And that "piercing" is certainly not the same as the "piercing" for a pencil or set of pencils functioning in the box—now used as a "pit" trap—where the pencils serve take on the role of vertical pungi stakes. Did we enter "holding" for the box corner, or "holding" for the pencil (now as an axe handle not a knife handle—again, two different forms of "holding")? These "functions" seem to be emerging dynamically under the analogy being contemplated (crossbow, beheader, pungi stake pit), or better, under the *transformation* we simply, conveniently, dub "analogy."

Perhaps we require greater generality. Instead of functions, we employ abstract "features" that seem to underlie the functions. For example, the "rigidity" of a pencil or of a spoon, the "flexibility" of the rubber bands, the "sharpness" of the razor. This was the route of a long-respected and well-respected cognitive science model of analogy (since we have entered momentarily into the subject of analogical transformations) by Gentner [Gentner1983], where analogies are formed by comparing (matching) these kinds of features. The connectionist model of analogy-making, DORA

1. **Prepare the cardboard box:**
   - Take the cardboard box and remove the flaps from one end to create an open space.
   - Cut out a rectangular section from one of the remaining flaps. This will serve as the crossbow arm.

2. **Attach the pencils:**
   - Attach two pencils vertically to the inside of the open end of the cardboard box. This will serve as the crossbow's frame.
   - Use the stapler and staples to secure the pencils to the cardboard box.

3. **Create the trigger mechanism:**
   - Take a toothpick and insert it horizontally through the crossbow arm, near one end.
   - Attach a rubberband to the toothpick on the inside of the crossbow arm.
   - Pull the rubberband back and hook it to the other end of the toothpick, creating tension.

4. **Assemble the crossbow:**

   . . . .

FIGURE 1.2. Part of ChatGPT's solution to the mousetrap problem (as of May, 2023).

(Discovery of Relations by Analogy) [Doumas2008] also followed this path of relying on mapping features to features. But this route too falls prey to the dynamics of transformations. A sock—dubbed "floppy" in the feature database—emerges as very sufficiently "rigid" when used in a fly-swatting transformation to swat a fly. The spoon, featuring the feature "rigid," suddenly has just enough flexibility to launch a pea at big sister across the kitchen table.

The generative pretrained transformers (GPTs) of generative AI, we shall see, choose a different route, in one way doing an end-around for the entire problem, in another way, simply following the same approach, but essentially relying on a form of a premade solution to the problem. In fact, a part of ChatGPT's solution to the mousetrap (as of May, 2023) is shown in Figure 1.2 where the GPT was told to build a mousetrap from the above-listed components "after the pattern of a crossbow." (Its "crossbow" purposely will not kill the mouse; only capture it.) Move a robot backwards in time (to include its camera and microphone), where the robot is guided in its actions via instructions from a GPT—back, back to the Stone Age. Remove, in other words, GPT's database of word-vectors derived from the vast knowledge of the human race as distilled in internet text. Place the robot and its GPT (or what is left of the GPT) in the pure ecological world. Now ask it to design a knife or build a rodent trap in an environment where the standard materials its robot tribe used are no longer available. This ecological setting, versus how the GPT can do this in our current world via the internet, is what is meant by the "end-around." This, of course will certainly be examined in the discussion to come.

TIME AND TRANSFORMATIONS

Transformations. These obviously are *creatures of time*; they occur over time, over what is at least—shall we say for the moment—*perceived* by humans as *continuous* time, as continuous change. The spoon stirring the coffee is seen continuously circling, the coffee's liquid surface continuously swirling. The sock employed in swatting is in continuous motion toward the fly. The rubber bands pulling the pencil back to its cocked position move in continuous motion. In this, curiously, we come to Roger Penrose and his infamous arguments [Penrose1994], inspired by Gödel's proof, for the necessary existence of what he termed "noncomputational thought." The arguments are infamous, we think we can say, because they were extremely adversely received, with much critique, by cognitive science and the AI community, with some shock to Penrose (see, for example, the comments on his précis of his earlier book [Penrose1990]).

Penrose envisioned a proof—a proof based entirely on visual transformations—of "a computation that does not stop" (i.e., an instance of the "halting" problem), to wit, each successive addition of a hexagonal number invariantly makes a cube (or cubical number). Therefore in a very commonsensical, mundane way, just as we might fold or bend the wire of a paperclip to function in some new version of a mousetrap, he portrayed a hexagonal number (e.g., 19) being *folded* into a three-sided, near-cubical structure (Figure 1.3), then *stacked* over the previous number (the previous number being a cube, 8), now making a new cubical number, 27. This folding/stacking/cubing process which is making successively larger cubes "obviously" (i.e., to the human mind) carries on infinitely. In this, we have simple operations of commonsense intelligence (folding, stacking) exalted to the status of participants in a mathematical proof, and one of Penrose's points was that *all* proofs rely on this level of knowledge, that is, on an "inexhaustible" pool of invariants derived from our commonsense experience—folding, stacking, cutting, pushing, stretching, and on.

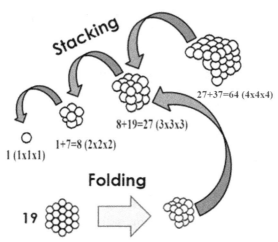

FIGURE 1.3. Penrose: A visual proof of a computation that does not stop—his *noncomputational* thought. A hexagonal number/form is folded (invariantly) into a three-sided cube, and stacked over the previous, invariantly making a cube. (Adapted from Penrose, 1994.)

Penrose was keying on this aspect being perceived (i.e., the always-resultant construction of a cube) as an invariance over this entire transformation of successive foldings and stackings, that is, an invariance seen over the *globality* of the transformation, over time, as a continuous event. This is what he thought required *consciousness*. What he was not explicit about, in fact, apparently did not explicitly see himself, is that he was simultaneously requiring some form of continuous time, again, at minimum at least *in our perception*, and equally an invariance perceived over this global, continuous transformation. This is to say, when he pointed to this form of "noncomputational" thought, a thought that requires consciousness, the "consciousness" he considered is inextricably linked to a perception over continuous time, in effect, to a form of *memory* "gluing" together as it were, as a whole extended in time, all the "instants" we like to think time is composed of, or in the computational framework, all the successive "states" of a computation. Without this glue, thus without the perception of the globality of the transformation, there is no perception of the invariance over the transformation also involved.

So, there are two things here: First, there is the source of the continuity of transformations, what we have already termed the *memory* or glue of the instants or states. Second, we are dealing with imagery—visual imagery, either in perception as we play concretely with mousetrap components in plain visual sight, transforming a pencil into a spear, rotating and positioning it into a hole in the side of the box, stretching/attaching the rubber bands—or *mental,* as we imagine the operations—folding, stacking—in things like Penrose's proofs.

Penrose was hardly the first to remark on this essential, transformational nature of thought. Virtually all the works of the great theorist of child cognitive development, Jean Piaget, could be said to be devoted to this, with titles such as *The Child's Conception of Space, The Child's Conception of Time, The Construction of Reality in the Child,* and *The Child's Conception of Movement and Speed.* One could also list Arnheim (*Visual Thinking,* 1969), Bruner (*Beyond the Information Given,* 1973), and Hanson (*Patterns of Discovery,* 1958). For one more illustration, let us take the gestalt psychologist, Max Wertheimer in his 1945 work, *Productive Thinking* [Wertheimer1945].

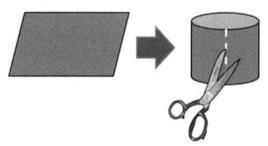

FIGURE 1.4. Computing the area of a parallelogram.

Wertheimer recounts being shocked by children in classrooms who were stuck, rigidly programmed by an algorithmic method for computing the area of a parallelogram, namely, the method of dropping a perpendicular from each of the top two angles of the figure. When he simply rotated the figure on the blackboard (with the normally horizontal edges now being vertical), the children exclaimed that they no longer understood how to compute the area. Yet Wertheimer had witnessed a five-year old girl with a cardboard parallelogram take scissors and cut off the two triangular ends, repositioning them to make a (easily area-computable) rectangle, and he watched another five-year old girl simply fold the cardboard parallelogram into a cylinder, then ask for a scissors so she could cut it into a rectangle—obviously both dynamic visual transformations of thought (Figure 1.4).

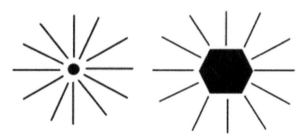

FIGURE 1.5. Left: Wertheimer's initial "point" intuition. Right: Wertheimer's "polygon" intuition. (Adapted from Wertheimer, 1945.)

Wertheimer recounted his own case of this when he was part of a restaurant conversation with an artist who asserted, "The sum of the angles of a polygon must always be 360 degrees." Wertheimer thought, "This is not possible, an isosceles triangle has 180 degrees ($3 \times 60°$), a rectangle has 360 degrees ($4 \times 90°$), and a hexagon has 720 degrees (6 × 120°)," but his mind started working on the problem, something he sensed to be correct. He recounts that the next day there came an intuition of a point, where the point is surrounded by an angular space of 360 degrees (Figure 1.5, Left) and he thought, "Is there something similar to this in a closed figure (Figure 1.5. Right)?" A few days later, the thought came, "Rather than a point, if I had a line (Figure 1.6, Top), it also has a space surrounding." And then, "How would I now proceed to obtain a closed figure?" His answer: "By *breaking* the line ... where the δ is the angle of rotation." (Figure 1.6, Bottom).

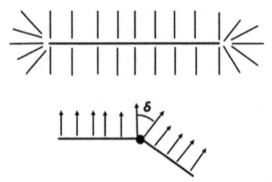

FIGURE 1.6. The intuition of breaking the line.
(Adapted from Wertheimer, 1945.)

And, if one continues this "breaking," the whole figure is formed as the line rotates and closes (Figure 1.7). The sum of the δ's must be, for a complete revolution, 360 degrees. And shrinking the figure to a point, we see the angular space made up of the δs.

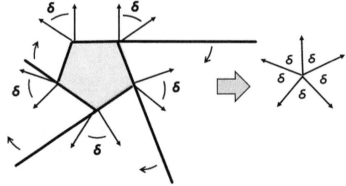

FIGURE 1.7. Rotating the line around to create the closed figure.
(Adapted from Wertheimer, 1945.)

Albeit with his "noncomputational thought" insight, it is obvious that Penrose is essentially restating Wertheimer along with many others in the older literature of cognitive science that existed before the advent of the "computer-equals-mind" cognitive revolution, all implying dynamic transformations preserving invariance over an indivisible, continuous flow of time.

THE IMAGE OF THINGS

The role, nay, even the need of imagery in thought has been an issue in cognitive science, while in AI, not an issue at all, for AI does not bother with a role for transforming imagery. Pylyshyn [Pylyshyn1973], a prominent cognitive theorist, only one year after Newell and Simon, simply denied any need of mental imagery, arguing that if images are simply constructed via the data elements and rules for manipulation of the data, the image is simply redundant – all the knowledge is in the rules and data structures in the first place: "In other words, the image has lost all its picture-like qualities and has become a data-structure meeting all the requirements of the form of representation set forth in earlier sections" (p. 22). Pylyshyn would go on to argue that the image can be put directly "… into one-to-one correspondence with a finite set of propositions" (p. 22), while "seeing the image" is now a matter of elementary logical, mechanical operations as in testing the identity of two symbols.

Thirty years later, Pylyshyn again challenged the cognitive science field to explain why images are needed [Pylyshyn2002]. The challenge he issued was in the form of a "null hypothesis." Although now admitting that his abstract symbol manipulations are likely insufficient to account for the form our representations take when experienced as imagery (as in Penrose's successively stacking cubes), he asked any future theory to explain, *why not*? Formal language and symbolic calculi, he noted, at least meet the dual requirements essential for reasoning, namely, *compositionality* (abstracted elements or concepts—shoe, pencil, hot, brown, paperclip, coffee …) and *systematicity* (lawful manipulation of the elements—"hot coffee," "brown shoe"). In contemplating certain "mental folding" experiments, where subjects were required to mentally fold paper into objects of certain forms, he noted that the subjects had, by necessity, to proceed sequentially through a series of folds to attain the result. Why? "Because," he argued, "we know what happens when we make a fold." It has to do with, he stated, "how one's knowledge of the effects of folding is organized."

The Pylyshyn of 1973, then, would simply deny a component of Penrose's proof and his argument for non-computational thought, namely, visual imagery as being important. The symbols, data structures, and logical manipulations account for it all, that is, *all* of Penrose's proof process. His 2002 version is simply agnostic. In this he would be joined by current AI which essentially committed early on to achieving *cognition without imagery*. The first component then—the continuity, the glue that holds all the states in a time-extended whole—would have to be applied to the operations of the computing machine, which, however, are always proceeding, *by definition*, state-by-state-by-state (i.e., by Turing's definition of computation), remembering that it is these *abstract operations that are critical*. Whether performed by a modern PC, a Turing machine with infinite tape, a register machine composed of beans and shoe boxes, or an abacus, the actual physical dynamics of these devices are irrelevant, only the effecting of the abstract operations is critical. This intrinsic discreteness of the operations, the missing "glue" of the states or instants, is itself is a problem, but the imagery problem—the question of whether dynamic visual imagery (or auditory, or kinesthetic) is intrinsically involved, in fact is required, for commonsense intelligence is not as easily disposed of as Pylyshyn or the AI field would wish.

There is the simple fact that we have already seen, namely, that we *perceive* the folding of things; we perceive the bending of things. Our visual experience of watching coffee stirring and using our hand in doing the stirring is where our knowledge comes from. Our visual perception of many a folding—of bed sheets, elbows, napkins, cardboard parallelograms—is where the knowledge of "how to make a fold," as Pylyshyn says, comes from. You can reduce this perceptual experience, as Pylyshyn did, to data structures, symbols, neural net weights, and propositions, but then there is no theory of how that perceived image of ours of the external world—the coffee cup and its spoon continuously in motion, stirring—arises from these sorts of "representational" structures, for surely, whether considered as bit configuration changes or neural flows, these look nothing like the coffee cup and stirring spoon—either within the computer or within the brain—and there is no known path other than the vague and frankly hopeful appeal to an equally vague "emergence" to explain the transition.

All this is to say that since cognitive science and AI have no theory of the origin of the dynamically changing image of the external world, there is no actual theory of *experience*, for example, the experience of seeing folds being made or liquids being stirred. Since there is no theory of the origin of experience, there can be no *grounded* theory of how this experience is *stored* or supposedly, as Pylyshyn envisions, how it is *organized* in the brain, especially if as lists of "finite propositions" (a "proposition set" which, given the structure of events via Gibson's invariance laws—like stirring the coffee—we shall see is simply not possible). Because we do not know how experience is stored, and since our experience is the foundation of our cognition, our entire theory of cognition is also ungrounded. It would certainly seem that it is crucial for both cognitive science and AI to derive a theory of the origin of the image of the external world.

THE IMAGE AS *SENTIENCE*

To be clear, this question of the origin and nature of the image of the external world is both the problem of *sentience* and consciousness. AI discussions like to take place around the question of whether an AI can be sentient, and better, whether an AI like GPT-3 may *already* be sentient, where sentience is defined roughly as, "able to perceive or feel things." The image of the external world is our experience is sentience. While we watch the coffee being stirred, liquid swirling, feeling the periodic motion of the spoon and the force being applied (applied to the spoon, and the resistance from the liquid), hearing the spoon's clinking, this is our image of the external world, or in other words, our experience. If this is not sentience, then we have no idea what to call it. Therefore, when we discuss the problem of the origin of the image of the external world, we are smack in the problem of sentience. Consciousness then supposedly gets added in, with a rough definition such as, "aware of and responding to one's surroundings." These two definitions are obviously heavily conflated—we are already clearly *aware* of the coffee swirling and stirring spoon as we perceive the event. And, neglecting the fact that we are already acting, doing the stirring (we do not have to stir in order to be sentient of the coffee being stirred), our ability to act upon the world is yet another question. Any surprise at the sentience problem being stated in terms of "the origin of the image" has its resonance in the strange, actually fatal philosophical misconception that somehow *perception* is not consciousness.[1] So, to repeat, a theory of the origin of the image of external world is crucial for cognitive science and AI, or rephrased for emphasis: *if we have no explanation of the origin of the image of the external world, we cannot possibly have an architecture that accounts for sentience.*

But how? The "origin of the image problem" is already well known, just in a slightly different form. It is known as the "hard problem." Chalmers stated it in roughly these terms: How, given any particular architecture, whether neural or computer, does that architecture account for the qualia of the perceived world [Chalmers1995]? Chalmers did not use the term "qualia" in his 1995 statement (rather, "quality"), but this is universally considered what was meant. The problem is that this statement of the problem has been a bad misdirection. The term "qualia" is usually exemplified with rather static examples: the red color of a sunset or of a rose, the taste of cauliflower, the brown and cream of the coffee surface. The question is then, how do such qualia arise from bit manipulations in computers or chemical flows in neurons? This has led to enormous philosophical effort to solve the problem in these terms, in fact the problem often being phrased, when reverting to Nagel's earlier formulation (and some of Chalmers own language), as explaining why an experience "feels like" something or "why it is like" something [Nagel1974]. In fact, perceived *form* is also qualia—spinning cubes versus slowly rotating cubes versus wobbly, plastically deforming cubes, buzzing flies versus flies slowly flapping their wings like herons. Note that this is already a function of the *scale of time* that the brain is tuned to *and creates* in its specification of the world.

---

[1] If the existence of this confusion seems questionable, view this paper ("Bergson, Perception and Gibson") in the *Journal of Consciousness Studies* [Robbins2000], then realize that the reviewers were reluctant to publish it due to concern that, "It is about perception, not consciousness."

The field of matter, or Minkowski's spacetime manifold if you will, has no particular scale of time—it can be imagined at many different scales. That usually visualized by the physicists, the scale of atoms and whirling electrons, is just one such scale, a "fly" now being a vague, effectively borderless cloud or ensemble of whirling electrons or quarks, or worse. "Buzzing" flies reflect our normal scale of time; heron-like flies would reflect a much slower scale. A heron-like fly cruising by the coffee cup would imply as well a much slower circling spoon, a much slower swirling of the coffee, and any verbal conversation at the kitchen table is going to be greatly slowed down, the syllables within each word far more spread out in time: con – ver – sa – tion. The "qualia" in the image, to include all its forms, are assuredly a function of a particular, specified-by-the-brain, scale of time. But these forms fully populate our time-scaled image of the external world and these forms are *dynamic forms*, continuously (that word again) changing over time. Here is Valerie Hardcastle's description of *qualia* [Hardcastle1995]: "… the conductor waving her hands, the musicians concentrating, patrons shifting in their seats, and the curtains gently and ever-so-slightly waving" (p. 1).

In other words, we have a qualia-filled image, full of dynamically changing forms. This is to say that the entire image of the external world is qualia. There is nothing in the image that is not qualia. The hard problem is better and more generally understood as this: *the origin of our image of the external world.* Yes, there is indeed an affective component in the problem: it is an aspect of the "why does it feel like something?" question. When I see the coffee being stirred, I might well think, "Yummm … coffee" or my spoon might feel a bit hot while I stir and the coffee liquid a bit sluggish to drive the spoon through. But the question of the origin of the image of the cup and stirring spoon "out there," on the kitchen table, has absolute priority, and in the framework within which we will see that the question must be answered, the nature of affect—at least its fundamental basis—naturally falls out.

Hinton, one of the great figures in neural net theory, is one AI-theorist who is aware that sentience is at least perceived to be a problem for AI, that it is equivalent to subjective experience, and that people think subjective experience is special to humans.

He notes that this is indeed GPT-4's assessment of the human attitude: "They [humans] think that the lack of subjective experience will prevent computers from ever having real understanding" (GPT-4, [Hinton2023]).

Aware of the problem's criticality to AI, Hinton's response is to downgrade it to zero in what he admits is akin to Dennett's approach to the subject (*Consciousness Explained*, [Dennett2017]). Hinton argues that people have completely misunderstood what the mind is and what subjective experience is, asking us to suppose that we have taken LSD, are now seeing pink elephants, and in wanting to describe what's going on in our perceptual field, we say, "I've got this subjective experience, like pink elephants in front of me." He argues that in examining what this means, we are trying to say what is going on in our perceptual system, but not by saying that Neuron #52 is highly active as that is useless. But we have this idea that things out there in the world give rise to percepts.

So, yes, before continuing Hinton's thought, "percepts," put more dynamically, would be our common experience of perceiving the coffee cup, swirling surface and stirring spoon, and the problem is explaining the origin of this dynamically changing, qualia-filled image. The previously explained part of Hinton's argument is as far as this question is going to be "addressed," no further—more on illusions, yes; veridical perception, no.

Hinton continues, stating that since we have this percept (the pink elephants), we would need to describe to someone what would be out there in the world for this to be the result of normal perception, and what would have to be out there in the world for this to be a normal perception is, indeed, a group of pink elephants running around. He notes that the subjective experience of pink elephants is not as though there is an inner theatre with pink elephants made of funny stuff called "qualia," but rather, we're trying to describe our perceptual system via the idea of normal perception, that is, what's going on here would be normal perception if there were pink elephants. Then, in perhaps the kicker for his argument, he states that what's funny about pink elephants is not that they're made of qualia in an inner world, but rather, that they're *counterfactual*—they're not in the real world, but they're the kind of things that could be. Thus, he argues, they're not made of spooky stuff in an inner theatre; rather, "... they're made of counterfactual stuff in a perfectly normal world, and this, finally, is what is meant when people talk about perceptual experience" [Hinton2023].

There is profound confusion here. The image of the coffee being stirred "out there" on the kitchen table is the "normal perception" to which Hinton refers. How did this image arise from neural chemical flows, manipulations of bits, neural nets firing? We hope, by this time, nearly thirty years since Chalmers' statement of problem, that projecting a matrix of pixels via a neural net for a homunculus to look at and see as a coherent image (not just a bunch of separate, independent pixels) is not considered a theory of experience. Nothing Hinton says answers Chalmers question: "How does the architecture account for this?" Is normal perception, as opposed to illusions, just *"factual stuff"*—this *mere phrase* explaining everything? Change the pink elephants to whirling, exploding coffee cups. These too came from "normal perception" or from some combination of experiences (via our memory) therein. With no theory of the origin of our normal perceptual images of coffee cups, the origin of the hallucinatory cup images is just as mysterious. Again, we shall see what a real model of the origin of the image—that "factual stuff"—looks like and why (as GPT-4's data model is indicating that millions of humans seem to believe) this question is indeed tied to real understanding.

THE CHALLENGE TO AI

AI, Hinton notwithstanding, in our awareness, has been rather blasé about the hard problem. In the way the problem has hitherto been stated and universally debated in the philosophical world, one can sympathize. As noted, since it is scarcely understood that perception (of the coffee stirring) is equally the problem of consciousness, this does not help. And staying within the standard (qualia) framework of the problem, one can ask, does it really matter if we can't quite state how the "redness of a rose" arises from our neural net architecture? What difference does this make to our getting on with language understanding, image categorization, causal reasoning, even plowing forward on that difficult problem of commonsense knowledge and AGI? But when the problem is reformulated as the origin of our image of a stirring spoon or of folding a sheet of cardboard, that is, the very *origin of our experience* of the world, thus the very basis of human cognition, to include how humans understand language, perhaps AI will reassess the problem's significance.

Rephrase the question: what is the nature of the "device" (be it brain, a different form of computer, something …) that can support and account for our image of the external world, in fact, a *time-scaled* image of the external material field—our normal, ecological environment? This will be the initial focus of this book. What we shall see is that the great French philosopher of mind, Henri Bergson, already in 1896 had stated a solution to the hard problem when the problem is indeed understood as the origin of the qualitative image of the external world [Bergson1896]. As the solution anticipated the essence of holography, many years before Gabor's 1947 discovery of the hologram, Bergson's framework was so prescient it was not understood by his contemporaries, nor ever since. To Bergson, we must join the theory of the great theorist of perception, J. J. Gibson, and his school of *ecological psychology*. Gibson is noted for his theory of *direct perception*—a concept little understood, but which becomes perfectly comprehensible when placed within Bergson's framework [Gibson1966], [Gibson1979]. Gibson is also noted for these things: his vision of perception as involving "affordances," the brain as in some sense engaged in a form of "resonance" (which Bergson will help explicate) and his emphasis on laws of invariance as defining the structure of external events, for example, events like stirring coffee, folding napkins, bending wires. In other words, in both the concept of affordances and the notion of events as structured by invariance laws, we find ourselves at the origin, in perception, of commonsense knowledge.

This basis, as founded in the nature of perception and in how the brain specifies the time-scaled image of the dynamically changing world, is going to demand a revision on how we think about memory, or better, how we think our experience is "stored," or per Pylyshyn (same difference), how the knowledge (experience) of folding is "organized," and how it is retrieved. This in turn, as indicated earlier, implies a different model of cognition and of how we construct those mousetraps and as well, a different model of the nature of human language understanding. What we will see is why, just like perception, *cognition requires consciousness*—a question with no answer in current cognitive theory, in fact, there being no understood need for consciousness (a bit disconcertingly for some) in the framework of the current computational conception of cognition which seems (we stress, *seems*) to work just fine without it. AI expert, Stuart Russell, in discussing the subject [Russell2019], simply notes that in the area of consciousness, we really do know nothing and that no one in AI is working on making machines conscious, nor would there be any idea where to start, and perhaps the critical point here, "*… and no behavior has consciousness as a prerequisite*" (p. 41, emphasis added).

Why is this going to be a challenge to AI? Because the framework of mind that is going to be laid out implies that both cognitive science as well as the philosophers of mind who monitor this subject can no longer be perfectly in league with AI, for hitherto cognitive science and most of philosophy have been nicely allied with the computational framework of mind. Chalmers devoted an article to the "Singularity" in a 2010 issue of the *Journal of Consciousness Studies* in which he very much supported this concept of AI in the future equaling, then exceeding, human intelligence [Chalmers2010]. The article was followed in the same journal (JCS, 2012, see [Chalmers2012]) by a set of twenty-six commentators. In his response to the comment-articles, by Chalmers' own count, fifteen totally endorsed or leaned strongly in his direction (the "Singularity" will happen), with another four merely neutral, making it 19/26 in at least implicit agreement. One of the detractors, Jesse Prinz (an AI theorist), turns out to feel we have already been simulated—by higher intelligences. Therefore the actual statistic could be 20/26. As far as the (six) strong detractors, Chalmers quickly disposed of them all. Francis Heylighen's critique [Heylighen2012], as an example, that intelligence cannot be treated as an isolated "brain in a vat," but must be *embodied* in dynamic interaction with the ecological world rather than isolated from that world, is summarily dispensed with: "As for embodiment and culture, insofar as these are crucial to intelligence, AI can simply build them in" ([Chalmers2012], p. 145). Perhaps, Chalmers allowed, this will require some time, but certainly within the framework of centuries.

But embodiment, we will see, cannot be dismissed nor so easily "built in," not within the conceptual framework in which AI currently works. The framework presented here will say that the brain is a much different device than AI currently envisions, that it is not software creation that is key, but rather engineering, engineering of a very concrete dynamics, as concrete a dynamics (although obviously far from the same) as that of an AC motor. We will see why this concrete dynamics is itself entirely embedded in, intrinsic to, the mechanism by which the coffee cup "out there" in the external world is specified, to include the physical/biological incorporation of Gibson's invariance laws structuring events. In regard to the latter, while theorists like John Searle have argued that it is the *biology* that is critical for consciousness [Searle2000], there has been little answer as to *why* this might be so, and AI has again somewhat justifiably ignored Searle's philosophical statement. But now there will be at least a partial answer to why a concrete dynamics is needed, although the extreme complexity of the very concrete, biochemical dynamics underlying this "device" can only be very partially explored, for this book only in terms of the (again, very concrete and complex) biochemical basis for the specification of the scale of time. There is however far, far more to explore in the biochemical substrate subject. All this is intrinsic to Gibson's concept of the (again, very concretely) "resonating" brain, resonating to an external event structured by invariance laws defined over time. This would require AI to now very seriously attend to ecological psychology for, as noted, this brings us to the basis of commonsense knowledge.[2] Therefore, AI will very seriously have to attend to the nature and implications of "embodiment."

But it is not just embodiment, for it is the physical, biochemical dynamics of our embodiment that is intrinsic to the solution of the hard problem when taken as we have construed it here—the origin of the image of the external world. Chalmers, whom we might appear to be singling out here, but simply because he is the most prominent and significant theorist in philosophy on AI, curiously seems to not understand the implications of the "hard problem" that he originally coined, even in (what we have termed here) its limited form, in fact, in recent discussions of what he terms "X" problems [Chalmers2022] that seem to be hurdles to AIs gaining *sentience*,

---

[2.] This will mean, however, that ecological psychology will also have to seriously review Gibson's framework, particularly in the light of his relation to Bergson. This book will equally be a "challenge" to current interpretations of Gibson.

which then presumably would mean the AI is actually having the *experience* (yes, thus the *image*) of stirring the coffee, the problem does not even merit mention as a low speedbump. This is reflective of the fact that Chalmers is no appreciator of concrete dynamics, one work for example [Chalmers1996], featuring a discussion wherein a "demon" is said to "preserve the dynamics" of a neural net by running around to all its nodes, noting inputs and outputs (connection weights) on a piece of paper (ultimately on stacks of paper). In other words, for Chalmers, it is just the symbols on the stacks of paper, that is, the purely symbolic manipulations, that are sufficient; concrete dynamics means nothing. How this would solve his original formulation in terms of "how any architecture (such as a neural net) accounts for qualia" is beyond us; he likely knows it does not, but the hard problem in its "just qualia" terms, has somehow faded (at least it seems so) in his mind into insignificance, an insignificance AI has been generally perfectly happy to accept.

But it cannot so fade; understanding the source of our dynamic, ever-changing image of the ecological world is essential; it absolutely underlies the theory of memory and cognition. Again, if we cannot explain the origin of experience, if we do not know what it actually is, then we cannot have a *grounded* theory of how this "experience" (of which we know nothing) is stored in the brain, and thus, how this experience is employed in cognition. But this—experience—is the source of human commonsense knowledge and "causal reasoning," in fact, as we shall see, equally the source of the "abductive" reasoning of C. S. Peirce, a missing element in AI's models, but part of an integral triad of induction-deduction-abduction (although, careful, for what abduction truly involves is not according to current AI conceptions thereof). Further, the qualitative nature of the image (just think of *one* case: the heron-like fly versus the "buzzing" fly) will be a severe problem for AI theory. Once this is seen, AI cannot "fade" the hard problem either.

Let us reemphasize this chain from the image to the commonsense knowledge (or AGI) problem, with the significance of understanding the essence of the problem in commonsense. AI theorists, in the context of the eventual "singularity," have taken to major worries and discussion about the "values alignment problem." How, worried Muehlhauser, if one asks an AI to, "get my mother out of the burning building," does one prevent the AI from simply blowing up the building, nicely ejecting your mom out of the building in the explosion [Muehlhauser2013]? How, worried Yudkowsky, does one stop an AI from "filling the well" as badly as the brooms do with their buckets when commanded to do so by Mickey, the sorcerer's apprentice [Yudkowsky2008]? But this "values alignment" problem is obviously simply the commonsense knowledge problem. If your super-intelligent AI does not understand that the *value* of a Toyota for stirring coffee is not all that high, or if it cannot realize a pencil has a nice *value* as an arrow in a crossbow mousetrap, its intelligence is not really going to be all that great, is it? It has hardly achieved AGI. The worry about "values alignment" only begins when the commonsense knowledge problem is solved: the two problems are the same problem, and that problem begins with the source of our image, thus our experience, of the external world.

Beneath all this lies a very fundamental problem—our conception of space and time. AI is built on a standard framework of our understanding of space and time, yes, a *metaphysic* of space and time. Physics, science, and especially mathematics are built upon this metaphysic. In fact, an intellectual evolution in physics occurred wherein a *mathematical* framework of explanation became accepted as, as equivalent to, a *physical* explanation. One can trace this evolution in the movement from Newton through Euler and Lagrange (the Lagrangian and Hamilton's "least action" being taken as the sufficient explanation of the physical paths of objects), unto Einstein and both special and general relativity. AI has been quite happy to ride along with this evolution in thought. In the Bergson-Gibson framework, this metaphysical framework no longer works; a new metaphysical framework—still very physical—involving a concrete dynamics is necessary, is required, to make the ecological model "go." This, we think, could be the most disturbing thing for AI.

But let us see what this vision of the brain, as a "device" supporting the specification of the image of the external world, looks like.

REFERENCES

[Bergson1896] Bergson, H. *Matter and Memory*. New York: Macmillan, 1896/1912.

[Chalmers1995] Chalmers, D. "Facing up to the problem of consciousness." *Journal of Consciousness Studies* (1995): 2(3), 200–219.

[Chalmers1996] Chalmers, D. J. *The Conscious Mind.* New York: Oxford University Press, 1996.

[Chalmers2010] Chalmers, D. "The singularity: A philosophical analysis," *Journal of Consciousness Studies* (2010): 17(9–10), 7–65.

[Chalmers2012] Chalmers, D. "The singularity: A reply to commentators" *Journal of Consciousness Studies* (2012): 8, 141–167.

[Chalmers2022] Chalmers, D. "Are Large Language Models Sentient?" October 13, 2022: https://www.youtube.com/watch?v=-BcuCmf00_Y.

[Dennett2017] Dennett, D. *Consciousness Explained.* New York: Little Brown and Co., 2017.

[Doumas2008] Doumas, L., Hummel, J., & Sandhofer, C. "A theory of the discovery and predication of relational concepts," *Psychological Review* (2008): 115(1), 1–43.

[Freeman1971] Freeman, P., & Newell, A. "A model for functional reasoning in design," *Second International Conference on Artificial Intelligence*, London, 1971.

[Gentner1983] Gentner, D. "Structure-mapping: A theoretical framework for analogy," *Cognitive Science* (1983): 7(2), 155–170.

[Gibson1966] Gibson, J. J. *The Senses Considered as Visual Systems.* Boston: Houghton-Mifflin, 1966.

[Gibson1979] Gibson, J. J. *The Ecological Approach to Visual Perception.* Boston: Houghton-Mifflin, 1979.

[Hardcastle1995] Hardcastle, V. G. *Locating consciousness.* Philadelphia: John Benjamins, 1995.

[Heylighen2012] Heylighen, F. "Brain in vat cannot break out," *Journal of Consciousness Studies* (2012): 19(1–2), 126–142.

[Hinton2023] Hinton, G. *"Two Paths to Intelligence,"* May 25, 2023: https://www.youtube.com/watch?v=rGgGOccMEiY

[Muehlhauser2013] Muehlhauser, L. *Facing the Intelligence Explosion.* Machine Intelligence Research Institute, 2013.

[Nagel1974] Nagel, T. "What is it like to be a bat?" *The Philosophical Review* (1974): 83(Oct.), 435–450.

[Penrose1990] Penrose, R. *"Precis of The Emperor's New Mind: Concerning computers, minds, and the laws of physics,"* Behavioral and Brain Sciences (1990): 13(2–3), 643–705.

[Penrose1994] Penrose, R. *Shadows of the Mind*, Oxford: Oxford University Press, 1994.

[Pylyshyn1973] Pylyshyn, Z., "What the mind's eye tells the mind's brain: A critique of mental imagery," *Psychological Bulletin* (1973): 80(1), 1–22.

[Pylyshyn2002] Pylyshyn, Z., "Mental imagery: In search of a theory," *Behavioral and Brain Sciences* (2002): 25(2), 157–238.

[Robbins2000] Robbins, S. E., "Bergson, perception and Gibson," *Journal of Consciousness Studies* (2000): 7(5), 23–45.

[Russell2019] Russell, S., *Human Compatible: Artificial Intelligence and the Problem of Control*, New York: Viking, 2019,

[Searle2000] Searle, J. R., "Consciousness, Free Action and the Brain," *Journal of Consciousness Studies* (2000): 7(10), 3–22.

[Wertheimer1945] Wertheimer, M. *Productive Thinking*. New York: Harper and Row, 1945.

[Yudkowsky2008] Yudkowsky,, E. "AI as a positive and negative factor in global risk." In N. Bostrom, M. Cirkovic (Eds.), *Global Catastrophic Risks*, Oxford University Press, 2008.

CHAPTER 2

GIBSON AND THE RESONATING BRAIN

BERGSON'S FORGOTTEN QUESTION

In 1896, Bergson published *Matter and Memory*. This was his driving question: *Is experience stored in the brain?* In his 1910 introduction to the book, he would state:

> Anyone who approaches, without preconceived ideas and on the firm ground of facts, the classical problem of the relations of soul and body, will soon see this as centering on the subject of memory ... ([Bergson1896], pp. xii–xiii)

For "the subject of memory," he meant exactly the above question: Is experience stored in the brain? For "the relations of soul and body," we can substitute "the relations of mind and matter." We can equally substitute "the hard problem of consciousness," for the problem—taken as the origin of our experience, or, as we prefer for corrective emphasis, the origin of our image of the coffee cup, swirling liquid, and stirring spoon—is surely sitting squarely in the problem of the relation of mind and matter. This is so whether the experience (or

FIGURE 2.1. Henri Bergson (1859–1941).

dynamically changing image) is taken as arising from either the matter of the brain or from the bit manipulations of a computer, or whether it is taken even in its limited form as accounting for the origin of *qualia* given some physical architecture, whether brain, neural net, or Turing Machine.

Before we go another inch into this "Is experience stored?" question, let us assure any already shocked reader that Bergson, before his 1896 publication, spent several years of study on the existing research on memory loss, particularly around aphasias and amnesia. He clearly held that *some form* of memory is indeed stored in the brain. This form is what today we might call *procedural* or motor memory. Such a memory would be the neural-resultant of piano practice session after session after session on Chopin's Waltz in C# minor; it is a form of memory that might be captured by a form of neural net and its gradual weight modifications. This neural structure allows us to unroll at will, as it were, the motor movements required to play that waltz. But all those separate practice sessions— the time when the teacher was furious with us, the time when the massive storm took down a tree outside the piano room—these are experiences. They are just as much *experiences* as sitting in the kitchen, stirring our coffee with a spoon. Again, we have the stirring with all its qualia, the storm with all its qualia, the angry, frustrated, fuming teacher with all that qualia. This is the question: Are these very hard-problem-esque, qualia-filled experiences stored in the brain?

To our knowledge, in the massive literature on the hard problem subsequent to Chalmers stating it in 1995, there has not been the slightest reference to, let alone treatment of, this "storage" question, whether in reference to Bergson's statement or just in general, that is, as seen as a problem by some consciousness theorist. This is an index as to how little the hard problem is understood. It is also of course an index into how deeply embedded is our concept of the brain as the "storehouse" of experience—how near to just a plain old *dogma* this is. But we are going to see why this question is critical in the problem of consciousness, and more precisely here, in conscious perception. We will look at a simple experience, stirring coffee with a spoon. What we will see is this: Any attempt to account for this ongoing, dynamically changing, conscious perception/experience of the coffee stirring via any model involving *storing* this experience in the brain (e.g., in "short term" or "immediate term" storage areas) is doomed to failure. That is, we are on the very ground floor of why the question, "Is experience stored in the brain?" is indeed critical to the problem of consciousness and to the hard problem. We are also at the core of another question: Is it possible for an AI, dwelling within its current computational metaphor, *to truly interact with the ecological world*, to achieve an experience of that world, in the way humans do?

STORING FEATURES—A CURRENT MEMORY MODEL

We are stirring coffee in a cup on the kitchen table, using a spoon. This is the simple experience we must account for via some concept of storing its structure—*as the event is ongoing*—in the brain. The seemingly inescapable propensity of current memory theory is to break this event into "features" that will be stored in various cortical areas. This was the approach to a memory model proposed by a very prominent set of memory researchers, Moscovitch, Cabeza, Winocur, and Nadel [Moscovitch2016]. Their assumptions were as follows:

1. The perception of an event: objects/scenes are comprised of "feature clusters."

2. Persistence (or continuity) of the event: There is persistence of the feature clusters during perception via neural activity.

3. Encoding the event (or a fraction thereof): The hippocampus becomes an "index" to multiple aspects of the event stored in multiple areas of the brain.

4. Retrieving the event: The HPC (hippocampal complex) index is engaged, retrieving and reassembling the event from the multiple aspects of the event which have been cortically stored in their respective areas.

Obviously, with its memory storage areas, the parsing into elements, the index for retrieval of the elements, the theorists envision a very computer-like process albeit in the flavor of the symbolic processing mode, and not so much, at least not so obviously, in the neural net mode at all, although both are going to have equivalent problems. But for the sake of exposition of the problem, we start with this model. Therefore, in this scheme, the ongoing, ever-changing coffee scene would be parsed into static features, for example:

- Feature 1: the cup's rim
- Feature 2: the cup's handle
- Feature 3: the top of the table
- Feature 4: the coffee liquid
- Feature 5: the spoon's handle
- Feature 6: the wood-grain of the table top
- And so on

There is no *principled* way—and this will become more apparent—to identify and list these features; the list is fairly arbitrary, could obviously continue at great length, could be far more finely detailed. It is up to the imagination of the theorist, but this is the essence of Assumption 1 (the scene is composed of features). Assumption 1 is also assuming that each of these features is being stored in some spot or specific area of the cortex.

For the sake of analysis in this event-as-features context, let us focus here on something apparently easy—the cup. Make it a whimsical, cubical cup with a handle, for surely a cube has features no one will take issue with—edges, vertices, sides. Yet Hummel and Biederman, in contemplating a connectionist model for object recognition [Hummel1992], already ran across a problem: If one breaks up the cube into these features (or breaks up some object, in their case a cube with a cone on the top face, Figure 2.2) and stores them in various cortical locations, how is Humpty Dumpty (so to speak of the cube) put back together again as a cube? One has to know what the original object looked like to reassemble the separately stored features properly, for example the vertices and the edges, in the right spatial arrangement.

Note here that reassembling is Assumption 4, and over the 4-D time-extended course of the event, for example, ten seconds worth of stirring, it means parsing/storing and reassembling at each instant or state (with its feature cluster) after state after state (with its feature

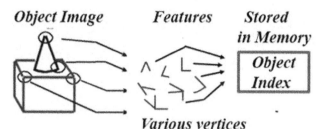

FIGURE 2.2. The features (deemed to be stored somewhere in areas of the cortex) can be reassembled in any number of ways—not necessarily as a cube and cone. (Adapted from Hummel and Biederman, 1992.)

cluster) in real time (choose the interval of sampling, perhaps, ten samples/second?) as some form of "representation" of each instant of the stirring event. This discrete time is an intrinsic attribute of the computer model in which these assumptions are embedded. Hummel and Biederman proposed decomposing the object into "geons" – elementary 3-D objects like cones, bricks, and so on. (to help keep the features together), but this model had its own problems [Robbins2004], to include essentially being an attempt to provide the brain with the nice picture on the cover of the puzzle box that helps us put the puzzle together.

But the situation is worse—we need only to start rotating our cubical cup. Now the use of features hits the *correspondence* problem [Adelson1985]. Imagine each state of a full rotation, for example, every one degree thereof, as a cartoon frame, thus one full rotation as a series (360 in number) of cartoon frames. As the cube rotates, frame by frame, each of the features (for example, the vertices) must tracked and the identity of each must be preserved across the frames (this is Feature X in frame 1, and this is the same Feature X in frame 2, and Feature X in frame 3, etc.) This problem was eventually declared *intractable* by perception theorists [Robbins2004]. This fact alone might have stopped Moscovitch et al. (and their assumptions) in their tracks.

VELOCITY FLOWS AND GIBSON

The theorists of form perception moved to *velocity flows* as key in the perception of form. This is where we can begin the discussion on J.J. Gibson and his ecological model of perception (Figure 2.3). Gibson's original insight was the mathematically structured nature of the ground that stretches out around us in every direction. This structure involved "texture density gradients" (Figure 2.4), where the ground consists of texture elements laid out in mathematical proportion. These are ubiquitous—gravel driveways, fields of grass, sands of a beach, a field of flowers. Turned upside down, the gradients become overhead ceilings stretching/narrowing into the distance or the stretching-in-the-distance of clouds. Stood up vertically on their sides, they become the walls of corridors stretching in the distance. For Gibson, the mathematical information in these gradients allowed them to be "specific to" a *surface* stretching in the distance. This was in direct opposition to Berkeley's concept wherein the points ABCD (Figure 2.4) all project to the exact same position of the retina, Berkeley thereby concluding that "distance of itself cannot be seen," and given this, we would need other information like tactile knowledge to see (i.e., to *infer*) distance. For Helmholtz too, and likewise for his modern exponent, Karl Friston [Friston2010], to see is a matter of inference. For Gibson, there is no inference taking place; there is this information defining the ground, thus being "specific to" distance, and it is preserved by a projective transformation on the retina varying lawfully with the movement of our eyes.

FIGURE 2.3. James J. Gibson (1904-1979).

A cup, moved across such a texture gradient (as would define the top of the kitchen table), maintains its size constancy in our perception by the fact that there is a constant ratio—an invariance—as it moves toward us or away from us: as the cup moves toward us, the number of texture rows (N) the cup occludes decreases in proportion as its height (S) grows, or $SN = k$. This is already an *invariance* law, expressing an invariant that is defined over time, that is, over the motion of the cup, such that our experience is of a cup of constant size moving toward us in continuous motion.

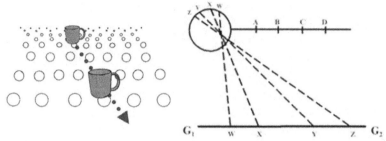

FIGURE 2.4. Left: Texture density gradient. The vertical separation (S) of the elements (across the rows) is $S \propto 1/d^2$, the horizontal separation is $S \propto 1/d$. The cup height (S) at the back occludes 4 rows (N), and this cup height doubles in moving to the front, but now occluding only 2 rows, that is, $SN = k$. Right: In Berkeley's conception, points ABCD project to the same point on the retina. For Gibson, the relative distances, wxyz are preserved by a projective transformation on the retina.

But such texture gradients are not static. We are constantly walking or driving or running across them. As such they become velocity flow fields (Figure 2.5), namely, an array of velocity vectors, the fastest moving vectors being closest to the eye of the moving observer, decreasing in velocity toward the horizon proportionally to the square of the distance from our observer—yet another invariance law.

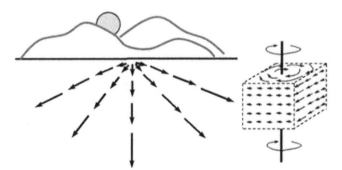

FIGURE 2.5. Left: An optical flow field—a gradient of velocity vectors, each with value, $v \propto 1/d^2$, where d is the distance from the observer. Right: A rotating Gibsonian cube, with an expanding flow field as a side turns towards the observer, a contracting field as it turns away, a radial flow field on the top surface.

So, theorists of form perception moved away from feature models with their intractable correspondence problem and moved instead to "energy models" [Adelson1985] and to Gibson's velocity flows as critical to specifying form. In doing so, they encountered the *aperture problem,* for the eye, being an aperture, occludes the actual ends of lines moving at some velocity (imagine looking through a circular aperture at a moving grating of lines whose ends extend beyond the borders of the aperture). This (occlusion) causes the computation of the direction/velocity of the lines to be uncertain [Weiss2002]. This led them to the position that within all perceptions there is an *intrinsic uncertainty,* that all perceptions must be viewed as probabilistic, a percept always being an *optimal percept* or optimal specification based on the best information possible. The brain then is forced to apply constraints on this probabilistic information. In the Weiss et al. model, the constraint, applied mathematically, is "motion is slow and smooth." Thus, a narrow, rigid, rotating ellipse (Figure 2.6) that eventually gets rotating a bit too fast, breaking this constraint, is now seen as a wobbly, *nonrigid* object (Mussati's illusion). This nonrigid object is now the optimal percept.

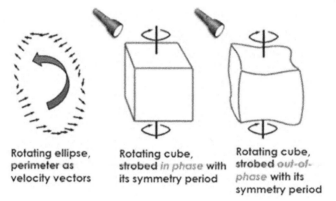

Rotating ellipse, perimeter as velocity vectors

Rotating cube, strobed *in phase* with its symmetry period

Rotating cube, strobed *out-of-phase* with its symmetry period

FIGURE 2.6. Left: A rotating ellipse, the perimeter a set of normal (right angle) velocity vectors. Right: A rigid rotating cube, strobed out-of-phase with its symmetry period, now a plastically deforming not-a-cube.

This optimal specification is a critical point, to be returned to, but another implication is also critical here: Consider what we'll term the "Gibsonian cube" (Figure 2.5). As a side rotates toward us, it presents an expanding velocity flow field. As the side rotates away, it presents a contracting flow field, while the cube's top is a radial flow field. What are the "features" in this cube—the edges, the vertices? These are now merely sharp discontinuities at the junctures of these flows. That is, the "features" are extremely dynamic, defined only over these flows. It is the flows that must be "stored."

The flows cannot be stored as a series of states (snapshots), each state holding a snapshot of the cube with all its features. The features are truly invariants defined over the flows, and if the flows are disrupted, strange things happen. An example of this can be found in the rotating cube made of wire edges (Figure 2.6, right). The cube has a symmetry period of four, being carried into (or mapped onto) itself every 90-degree rotation or four times per full revolution. If strobed at an integral multiple of this symmetry period—perhaps, 4, 8, or 12 times per revolution—the cube is seen as a rigid, rotating cube. If strobed out-of-phase with this period, at a nonintegral multiple, for example, 5, 9, or 13 times per full revolution, the rigid "features" of the cube disappear—it becomes a plastically deforming, nonrigid object, and not a rotating, rigid cube [Shaw1974].

The first thing to note in this is that each strobe is a *sample*, a snapshot, and the features (vertices, edges), if indeed statically defined, should exist in this snapshot or sample. But to the brain, they apparently do not so exist. Second, the brain, again, appears to be expecting a certain constraint to be fulfilled (or simply, an environmental statistical regularity), and just a guess on this: *spatial symmetry* (as in a regular cube) implies *temporal symmetry* (regular, periodic pulses "emitted" as it were by each side as a symmetric, very regular cube would do when rotating). If not met, the "optimal specification" is to a plastically deforming, wobbly, nonrigid object. The brain is specifying (but still optimally) deformed flows.[1]

If we recur back to our starting model, namely, that of identifying features within the incoming perceptual signal, storing them within some short-term memory area, then reassembling these in real time as the perception (or "representation") of the cubical coffee cup (which we set rotating), the coherence of it all is dissolving. The concept implies:

- We could track the features from stored state to stored state as some mechanism reassembles the event, that is, the intractable correspondence problem is denied as being a problem at all.

---

[1] Given the many possible rotation speeds, the brain would require a form of precognition to apply the right in-phase sampling frequency to always see a "cube," and as Turvey [Turvey1977] noted, what if we had two or three cubes rotating at different speeds simultaneously?

- It implies that there are plenty of hidden puzzle-box covers sitting around to aid the brain in putting the features of the cup (or some other object) back together properly in the first place.

- It ignores that these features seem to exist only over flows, but how are flows "stored," if not as a series of states (i.e., sampled, static images of the cube)?

In fact, in bringing this last point up, we realize that the entire reassembling concept—sets of features assembled and re-presented in state after state after state – is *surreptitiously assuming a fundamental continuity*, in essence, it is *assuming consciousness* itself, for how is the series of reassembled states seen as a continuity— as a flow, as the continuous "rotation" of the cup? How is it (i.e., our experience, per this model) not just a state, then another state, then—another state, that is, never more than *one state*, one "instant," one "slice" of the rotation of the cube (i.e., its reassembled, static features)? In essence, it is equally assuming *memory*—that which provides continuity beneath the states, the mysterious "glue" of the states. In fact, we are getting a glimpse of the concept that memory and consciousness, as Bergson saw, are one and the same problem.

The preceding statement is just a different expression of something noted by Turvey [Turvey1977]: If state (or snapshot) after instantaneous state of the coffee stirring must be stored in a static space (in the brain) like a series of snapshots laid out on a desktop—a vast, static structure, how does one reintroduce the motion? As Turvey pointed out, via some "internal scanner"? Then how, he argued, does the *scanner* register motion? An infinite regress begins.

If we pause here to address these issues taken rather in the neural net (and/or deep learning) framework, not that much changes. The feature problem seems to go away: the network can be trained to recognize a certain class of images from a set of images and consistently assign it a value indexed to a label, for example, the label, "stirring a cup of coffee." The arbitrary theorist-created list of features of the event is gone, the net having defined its own limited set of features, a set difficult (via "peering" into the net) to exactly understand or determine (in essence, pixel groups). In other words, there is still a form of static features, but now stored in the distributed weighting learned and defined throughout the net. And, note, this is a limited set; it is not in any way the full set of features that could possibly be defined over any given event such as our coffee stirring. This will be important later.

What then about *recognizing* a dynamic, ongoing event of coffee stirring? Although it is doubtful that in the training on various static images, the net would have made any distinction between images with the spoon in different positions (along with the liquid's swirl in different forms, stages, phases), classifying them all as "stirring coffee" (assuming all were in the right ball park—that all had the cup, liquid and spoon at least in a stirring position), it is safe to say that any algorithmic route will involve training the net, even if in the ostensible form of videos, on a series of static images (e.g., with the spoon in different positions in the cup). Even if a video, it is still being presented as a set of frames or snapshots. In any case, the net is simply unreflective of the concrete continuity of stirring. Add in a robotic arm computing the arm/hand's next 3-D coordinate position after 3-D coordinate position, each position linked to an image in the series—this lack of concrete, experienced continuity does not change.

But this leads us to a worse problem, namely, the very structure of this coffee stirring event that is also being ignored.

STORING EVENT INVARIANCE STRUCTURES?

The stirring of the coffee is an event described by a set of invariance laws—a concept we have already seen in the Gibson framework, and which is central. These laws generally are explored by the Gibson school of ecological psychology. We've seen one such in the velocity flow field with its velocity vectors proportional to $1/d^2$ from the eye of the observer. A partial list of the laws involved in our coffee stirring is as such:

- a radial flow field defined over the swirling liquid
- an adiabatic invariant carried in the haptic flow re the hand-arm and spoon, that is, a ratio of energy of oscillation to frequency of oscillation [Kugler1987]

- an inertial tensor defining the various momenta of the spoon, capturing the object's resistance to rotational acceleration in different directions, and specific to the object (spoon) being perceived as having a constant length and width [Turvey1995] [Turvey2011]
- acoustical invariants
- constant ratios relative to texture gradients and flows for the form and size constancy of the cup
- a ratio (termed *tau*, to be discussed in the following paragraph) related to our grasping of the cup [Savelsbergh1991]
- and more …

So, the (ongoing) event has a time-extended invariance structure. The swirling coffee surface is a radial flow field, while the constant size of the cup, as one's head moves forward or backward, is specified, over time, by a constant ratio of height to the occluded texture units of the table surface gradient. Over this flow field and its velocity vectors there is the value, τ, (tau), specific to the time to impending contact with an object or surface, and with a critical role in controlling action. As the coffee cup is moved over the table toward us, this value specifies time to contact and provides information for modulating the hand to grasp the cup. As the cup is cubical, its edges and vertices are sharp discontinuities in the velocity flows of its sides as our eyes saccade, where these flows specify, over time, the form of the cup. The periodic motion of the spoon is a haptic flow field that carries an adiabatic invariance—a constant ratio of energy of oscillation to frequency of oscillation, and another aspect to this motion of the spoon is a form of "wielding," described under the concept of the "inertia tensor."

These invariants are simultaneously multimodal and *amodal*, as they define the *same information* (amodal) in multiple modalities: the periodic motion of the spoon is precisely coordinate with the visual flow field and with the acoustic "clinking" as spoon collides with the cup's side. (A Gibson example was that of running one's finger down the teeth of comb, where the successive staccato sounds are coordinate with the visual "wave" of teeth, bending/snapping back, passing down the comb [Gibson1966].) There is no need for solving the "binding problem," binding these aspects of the event together as a whole—they are intrinsically bound as coordinate, dynamic information. The information is *coherent*. Finally, this entire structure and far more must be supported, globally, over time, by the resonant feedback among visual, motor, auditory, even prefrontal areas. In other words, it is this entire informational structure that must be supported, in ongoing fashion, by the neural dynamics supporting the perception of the coffee stirring event.

These invariants—like the "edges" and "vertices" of the rotating cube—are defined only *over time*, over the flowing transformation of the external field of which the coffee stirring is but a subset. These— for example, the adiabatic ratio, the inertial tensor—do not exist in a static instant. So, if we are disassembling the event into "features" in real time and storing these in a very short-term memory, and of course, reassembling these in real time as required for the needed, usually assumed internal or mental "re-presentation" of the stirring event, we must wonder where, *how and even when* is this structure of invariance with its requisite flows being stored (let alone what marvelous algorithm is driving the selection of all the unique cortical locations into which to store these features during the event, and for every event, to chemically "consolidate" the storage for some unspecifiable period, even for *years*)?

All coherence with respect to the disassembling or parsing into static "features," storing these, then reassembling them, is again dissolving here. The adiabatic invariant may be dubbed a "feature," but it is not extractable and storable as a static element. And again, even if one could make sense of placing this "feature" successively in a series of "states," where is the continuity coming from—the "time glue" so to speak—for what Kugler and Turvey saw as a *haptic flow field* with its adiabatic invariance [Kugler1987], itself part of the continuous, periodic motion of the stirring spoon?

It is perhaps easier to highlight the many lurking problems if we simply look at things from another possible point of view. What if, we can ask, the brain is simply *reflecting* this dynamically changing invariance structure defining the coffee stirring, symmetrically, in near simultaneity with the event (the "near-simultaneity" being the allowance for the slight time delay needed for transmission of the information picked up in the optic array and then to/through the neural structure)? This is to say, there is no attempt happening at "storage" in (some yet to be unequivocally found) short-term storage areas, and this is simply because, as one can see in contemplating the nature of these dynamically changing invariance structures with their flows, there really can be no such attempt—*there simply are no logical points in time at which to be storing anything!*

SYMMETRY, RESONANCE: NO STORAGE

This was Gibson's position [Gibson1966]. He envisioned the brain "resonating" to the external event, something elucidated more in his description of *perceptual activity* wherein he envisions the brain continuously adjusting, tuning to the invariants, and seeking the highest fidelity reception relative to the (dynamic) information. Symmetrically, then, the invariants specifying the event are equally defined over this resonant, attuning activity of and in the brain. Thus, while Gibson is noted for arguing that the information in the external world (the invariants) is "specific to" (or specifies) the external environment, for example, the *texture density gradient* is specific to a surface stretching in the distance (as in Figure 2.4), it is true by symmetry to also say that the dynamic state of the brain "is specific to" the external event.

An example we can take is the *tau value*, a ratio defined over a velocity flow field that is specific to *severity of impending contact* (Figure 2.7), as we've already mentioned [Lee2009]. The value of this ratio is critical for controlling landings if you are a flying bird descending to land on the ground or if you are a pilot landing a plane. This dynamically defined value, tau, is considered to be reflected in the action systems of the bird, guiding the modulation of its wings. In a word, this value defined over the flowing field is seen as necessarily reflected, symmetrically, in the resonating processes of the brain.[2]

[2] One might question immediately, wouldn't this sort of information be employed in an automatic driving system, perhaps, to slam on the brakes if approaching the rear of another car too quickly? We are not privy to the exact algorithms being used in these applications, but two points need to be made here: First, we're interested here in creating a clear understanding of the kind of information the human brain/body is using in interacting with the ecological world. Second, even should an auto-driving AI happen to be computing a tau value, it would necessarily do so via time sampling and memory storage, and in this, there would not be the *experience* that, for example, a bird has in feeling itself approaching too quickly in relation to the flow of the "landing field," having

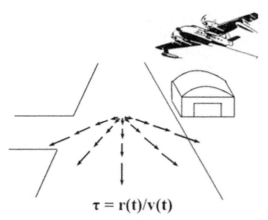

$$\tau = r(t)/v(t)$$

FIGURE 2.7. The *tau* value is specific to *severity of impending impact*, where tau is defined as the time derivative (tau) of the inverse of the relative rate of optical expansion.

Raja recently emphasized this symmetry [Raja2018], noting that this fills in more fully the nature of Gibson's "resonance" description of the brain, that is, where the brain is simply "resonating to" the coffee stirring, and not storing things like features. Raja used, as one of many examples, the well-studied (by ecological psychology) invariance laws employed by a baseball outfielder to guide his action systems to intercept and catch a fly ball. In this case, the fielder constantly moves in order to cancel the lateral motion and vertical acceleration of the fly ball relative to his own movement so as to maintain a homogenous expansion of the ball (as an expanding flow field) in his visual field. Raja noted that, "… resonance is what is going on inside the organism, especially in the central nervous system (CNS), with regard to what is going on at the ecological scale" (p. 33). Thus, the invariants specific to the baseball's arc through the air (and simultaneously to the action-systems involved in the outfielder's catching the ball) must be reflected in the neural processes of the brain.

Raja employed a description of the CNS and the invariants of the environmental event based in dynamic systems theory (DST), an approach which, given the right parameters describing the attractor patterns, demonstrates this symmetric reflection of the invariants.

---

to modulate its wings to slow down and perhaps brace its legs for the coming crash. And this latter (the time-extended experience) is going to be the absolute core of the issue.

Such a view of things removes the clunky approach of storing features of the coffee stirring or of the flying baseball (whatever such features would be in this case!), instant by instant, ridden as this is with logical dilemmas. It still leaves a memory dilemma.

A DIAGRAMMATIC GAP: THE PHENOMENAL PAST

In this symmetry schema, where the brain is simply resonating to and with the structure of the event, we nevertheless noted it would be only a *near-simultaneous* symmetry. The event (ultimately perceived) is still in the *past*; the event is a limited extent in time of what was a *past* transformation of the external material field. Facing the coffee cup, we are perceiving the position of the spoon nearing the 3 o'clock position as it is/was moving from the 12 o'clock position. But this "present" perception is already of the *past*—more optical information is arriving at the retina as the spoon is happily on its way toward the 6 o'clock position. This may not seem a problem if we think the image/perception of the coffee cup is simply "in the head," or presented in some "mental space" or "perceptual space," but these vague concepts, no longer involving the physical world, are just convenient hypotheses to put the perception "somewhere." This question we will return to, but the accompanying problem is that we are seeing the *motion* of the spoon—some time-extent of its movement from, for example, the 12 o'clock to the 3 o'clock position on its circle. We are back to the memory underlying this, something we have seen has great difficulty being a series of stored "states."

Even the resonance view would seem then to share something akin to the second assumption of Moscovitch et al., namely, "There is persistence of the feature clusters during perception via *neural activity*." Taylor [Taylor2002] used a similar concept, arguing that the features of an object are bound to "activity in working memory," with this providing our experience (the content of consciousness) of the object, for example, the coffee cup. He saw this neural activity "relaxing" to a temporally stable state, with this providing *"the extended temporal duration of activity necessary* for consciousness" ([Taylor2002], p. 11).

What is the time-extent of this neural activity? Just long enough to specify the spoon in motion from the 12 o'clock to 3 o'clock position, and from 3 to 6 (allowing for the sake of discussion that there is a hypothetical sliding "three clock-position window" moving from 12 to 6)? Whatever the theorist needs? By what principle is this neural activity limited to a certain extent in time? Why does it not extend to the entire past? Why is the brain not a vastly extended 4-D object per this conception with no need for storing anything? The fact is that the "extended temporal duration of activity necessary for consciousness" is merely a convenient device to slip in the continuity needed for perception, that is, it is a surreptitious way to slip in an explanation for consciousness, simultaneously, to slip in an explanation for the memory underlying consciousness, for inescapably, consciousness implies memory.

At this point, we think we have seen pretty clearly why, taken in the basic, elementary context of the perception/experience of stirring a cup of coffee, the question as to whether our experience is indeed stored in the brain is critical, for we discover that it is very difficult to even begin to explain our perception—our *sentience*—of a simple event like coffee stirring when relying, implicitly or explicitly, on this instant-by-instant, storing-in-a-memory framework. This is equally to say that the question is a critical component of the hard problem, for surely, just taking this problem in its usual, although limited and misdirecting terms of "accounting for qualia," the stirring itself, the aroma, the clinking, even the *forms* as we saw with the rotating cube,

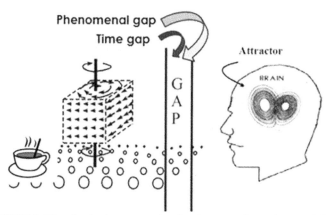

FIGURE 2.8. There is a both a temporal and a phenomenal gap: the event is in the past; the external event looks nothing like the "attractor" (or the chemcial flows comprising an attractor) in the brain.

are all *qualities* arising, changing, building over time, thus needing that memory, exactly as we saw Valerie Hardcastle envisioning qualia in terms of motions and changing forms, with the orchestra's musicians concentrating, the patrons shifting in their seats, and the curtains gently and ever-so-slightly waving.

But that "attractor" that Raja envisioned being supported by the resonating brain looks nothing like the rotating cube or the coffee being stirred with all its qualia of form, color, acoustics, and kinesthetics, or, for that matter, anything like Hardcastle's rustling orchestra. In other terms, the neurochemical flows that make up this attractor look nothing like the coffee and circling spoon. There is a two-fold gap—both of *time* (the event is in the past) and the *phenomenal* (Figure 2.8). The qualitative image of the external world—our fundamental form of experience, our *sentience*—is unaccounted for. This is the problem we turn to next—resolving the gap. We will see that the two aspects of the gap—time and phenomenology—are one and the same problem.

REFERENCES

[Adelson1985] Adelson, E., & Bergen," J. "Spatiotemporal energy model of the perception of motion," *Journal of the Optical Society of America* (1985): 2(2), 284–299.

[Bergson1896] Bergson, H. *Matter and Memory*. New York: Macmillan, 1896/1912.

[Friston2010] Friston, K. "The free-energy principle: a unified brain theory?" *Nature Reviews Neuroscience* (2010): 11(2), 127–138.

[Gibson1966] Gibson, J. J. *The Senses Considered as Visual Systems*. Boston: Houghton-Mifflin, 1966.

[Hummel1992] Hummel, J. E., & Biederman, I. "Dynamic binding in a neural network for shape recognition," *Psychological Review* (1992): 99(3), 487–519.

[Kugler1987] Kugler, P., & Turvey, M. *Information, Natural Law, and the Self-Assembly of Rhythmic Movement*. Hillsdale, NJ: Erlbaum, 1987.

[Lee2009] Lee, D. N. "General Tau theory: Evolution to date. [Special Issue]: Landmarks in perception," *Perception* (2009): *38*(6), 837–858.

[Moscovitch2016] Moscovitch, M., Cabeza, R., Winocur, G., & Nadel, L. "Episodic memory and beyond: The hippocampus and neocortex in transformation," *Annual Review of Psychology* (2016): *67*, 105–134.

[Raja2018] Raja, V. "A theory of resonance: Towards an ecological cognitive architecture". *Minds and Machines* (2018): *28*(13), 29–51.

[Robbins2004] Robbins, S. E. "On time, memory and dynamic form," *Consciousness and Cognition* (2004): *13*(4), 762–788.

[Savelsbergh1991] Savelsbergh, G. J. P., Whiting, H.T., & Bootsma, R. J. "Grasping tau," Journal *of Experimental Psychology: Human Perception and Performance* (1991): *17*(2), 315–322.

[Shaw1974] Shaw, R.E., & McIntyre, M. "The algoristic foundations of cognitive psychology" In D. Palermo & W. Weimer (Eds.), *Cognition and the Symbolic Processes,* Hillsdale, NJ: Erlbaum, 1974.

[Taylor2002] Taylor, J. G. "From matter to mind," *Journal of Consciousness Studies* (2002): *9*(4), 3–22.

[Turvey1977] Turvey, M. "Contrasting orientations to the theory of visual information processing," *Psychological Review* (1977): *84*(1), 67–88.

[Turvey1995] Turvey, M., & Carello, C. "Dynamic touch," in W. Epstein & S. Rogers (Eds.), *Perception of Space and Motion*, San Diego: Academic Press, 1995.

[Turvey2011] Turvey, M., & Carello, C. "Obtaining information by dynamic (effortful) touching," *Philosophical Transactions of the Royal Society B: Biological Sciences* (2011); *366*(1581), 3123–3132.

[Weiss2002] Weiss, Y., Simoncelli, E., & Adelson, E. "Motion illusions as optimal percepts," *Nature Neuroscience* (2002): *5*(6), 598–604.

CHAPTER 3

BERGSON AND THE IMAGE OF THE EXTERNAL WORLD

BERGSON, TIME, AND DIRECT PERCEPTION

For Gibson, perception is *direct*. This is to say that there are no internal representations, no "internal models" of the external world in the brain, no internal "theatre" (with a homunculus viewing the image). The image of the coffee cup is not "in the brain," or within its neural structure somehow, somewhere, or worse, in some ill-defined, mysterious "perceptual space" or "mental space." This latter conception of internal representations of a very real, objective world out there, that is, assuming a form of realism about a truly objective world, is *indirect* realism. By direct perception (or direct realism) is meant that the coffee cup is perceived exactly "where it says it is," out there, on the kitchen table, without any internal, representational intermediaries. For Gibson, the information in the environment, for example, the texture density gradient, *directly specifies* an external ground expanding in the distance. Again, perception is direct.

For many in the Gibson school, this is already enough— we need add no more to explain how the brain is specific to the environment or in the more standard statement, how invariants are "specific to," in fact, *unambiguously* specific to, the environment, that is, how we see the coffee being stirred "out there," on the table. Turvey [Turvey2019], noting how the animal both sees the surface and *walks* on the surface, states simply: "There is no other 'object' between the animal and surface. This two-term relation is all that is meant by direct perception" (p. 28). Relatedly, "To perceive things is to perceive how to get about among them and what to do or not to do with them" (p. 365). The latter, however, gives us little basis to distinguish from a robot (also navigating among things) and theorists of consciousness, focusing on the brain and/or the robot architecture, will repeat Chalmers' question: how does this architecture account for qualia (by implication part of the specification of the world)? And unfortunately, a principle like "this two-term relation is ... direct perception" is a level of abstraction that is yet hardly comprehensible. Lehar [Lehar2001], in his discussion of Gibson's statements on direct perception can make no sense of it. Purves and Lotto [Purves2010] flatly declare that Gibson's direct perception is generally viewed as "mystical." Searle [Searle2015], actually trying to argue for and build a theory of direct perception, never bothers to even mention Gibson! Something is missing for direct perception's comprehension.

The brain in this time-extended resonance metaphor, with its perceptual activity, is still a mass of neural flows, where something about these flows and their chemical velocities is establishing a proportionality or ratio—a ratio of "brain events," shall we say, to the elementary micro-events of the external field, say, to the micro-events of the fly moving by, such that a scale of time is specified—the "buzzing" fly or our normal scale. Again, the surrounding environment, to include the fly going by, in the physics world is the time-scale-less Minkowski manifold, something which, taken in the physics story, looks nothing like flies, coffee cups, or coffee. Further, there is obviously no image of the coffee cup or of the fly identifiable in this mass of neurochemical flows. Addressing this aspect of the problem, we say that the flows (supporting a resonance) are "specific to" the coffee stirring, that is, to the set of invariants defining these events (or the flows are in effect "resonating to" this structure). Nevertheless, "specific to" is a *magical* term here. It carries no actual mechanical, physical, concrete meaning that allows one to make concrete sense of it, that is, to concretely understand how there is now an *image* of the world (at some scale of time). It is yet a level of vague abstraction, but this really should not be so.

There is a *concrete* example—a very physical process—of specification of an image [Robbins2000], [Robbins2006]. It originated with Bergson. It appears in the first chapter of *Matter*

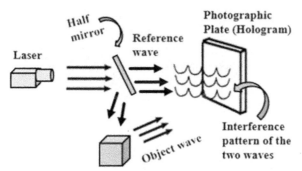

FIGURE 3.1. Constructing a hologram. Each point of the illuminated object/scene, for example, a point on the cube, reflects a spreading, spherical wave towards the plate, each such wave covering the entire hologram plate. This composite "object wave" from all the points forms an interference pattern at the plate with the reference wave from the laser. The information for each point of the object then is found everywhere on the hologram plate, and conversely, *the information for the entire object is found at each point on the hologram plate.*

and Memory (1896), a chapter considered "obscure" by his contemporaries. In it, Bergson noted that there can be nothing like a "photograph" of the external world developed in the brain. We will find nothing remotely looking like the coffee cup and spoon inside the skull. The neuroscience of his day was already clear enough on this. But Bergson went on:

> But is it not obvious that *the photograph, if photograph there be, is already taken, already developed in the very heart of things and at all points in space.* No metaphysics, no physics can escape this conclusion. Build up the universe with atoms: Each of them is subject to the action, variable in quantity and quality according to the distance exerted on it by all material atoms. Bring in Faraday's centers of force: The lines of force emitted in every direction from every center bring to bear upon each the influence of the whole material world. Call up the Leibnizian monads: Each is the mirror of the universe ([Bergson1896], pp. 31–32, emphasis added).

This was Bergson's declaration, 51 years before Gabor's 1947 discovery of holography, and 80+ years before Bohm [Bohm1980], that the universe—yes, our environment—is a hologram (Figure 3.1) or better, a holographic field, and that at every point in the universe is the information for—the "photograph" of—the whole. But unlike Pribram [Pribram1971] where the *brain* is the hologram, or Bohm (1980) where the brain, very vaguely, somehow is involved in unfolding an "explicate" order in the holographic field, we can say (in updated terms) that Bergson saw the brain as being (or creating, or supporting) a *modulated reconstructive wave* passing through this holographic field (Figure 3.2), specific to a source within the field. The neural processes—action potentials, neural spikes, and so on—seen currently as supporting "computations," in fact all that is presumed to support the "resonating" brain, become integral participants in the formation of this concrete waveform (Figure 3.3). Yes, this is obviously a very complex wave form, far more than a simple, coherent "frequency."

Immediately after the "photograph" passage, Bergson noted:

> Only if when we consider any other given place in the universe, we can regard the action of all matter as passing through it without resistance and without loss, and the photograph of the whole as translucent: Here there is wanting behind the plate the black screen on which the image could be shown. Our "zones of indetermination" [organisms] play in some sort of the part of that screen. They add nothing to what is there; they effect merely this: That the real action passes through, the virtual action remains ([Bergson1896], p. 32).

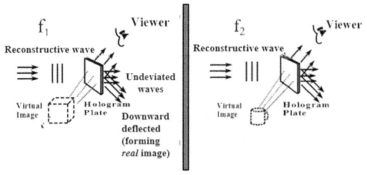

FIGURE 3.2. Modulating the reconstructive wave. Two object wave fronts were stored on the same hologram plate, one using a reference wave of frequency 1, the other, frequency 2. Modulating the *reconstructive wave* to frequency 1 specifies the cube; frequency 2 specifies the cup.

In Bergson's terms, the universal field is a vast field of "real actions" (one can read "waves" for concreteness) rippling everywhere—a vast interference pattern. Any given "object" acts upon all other objects in the field and is in turn acted upon by all other objects. It is in fact obliged:

> ... to transmit the whole of what it receives, to oppose every action with an equal and contrary reaction, to be, in short, merely the road by which pass, in every direction the modifications, or what can be termed real actions propagated throughout the immensity of the entire universe ([Bergson1896], p. 28).

The subset of these actions (or information) that the brain-supported reconstructive wave picks out is a portion related (or relatable) to the body's action capability. This action-relatability is the information-selection principle from the "hologram." Thus, perception, as Bergson argued, is *virtual action*. We are seeing how we can act. If put in a more Gibson-like, ecologically familiar way, we are seeing what the environment "affords." An *affordance* in the ecological school is defined as a relational property that supports the action of a certain organism, for example, properties of a surface layout that are specific to a possible action such as the texture gradient that specifies it is "walk-on-able" or if a snake, slither-on-able. The relational aspect is brought home by examples such as the *pi ratio*—a ratio of stair riser height to knee height, which—as long as within a certain value—specifies that the stairs are "climbable" for a human, and beyond that value, not-climbable [Warren1984]. As the brain's reconstructive wave-specification must be placing a certain *scale of time* upon the external field, for example, the normally seen wings-ablur, "buzzing" fly versus a fly slowly flapping his wings like a heron, we shall see the virtual actual principle, and equivalently, the affordance principle, have definite time-scale implications for action.

In this, then, the image is not mysteriously "emerging" from the brain's processes; the brain is not mysteriously "generating" an image; it is not generating "experience," and there is no image in some mysterious "perceptual space" or "mental space." The image—the coffee cup and stirring spoon "out there"—as a specification of a dynamically changing, time-extended *past* subset or aspect of the field, is within the external field, right as commonsense says, "where it says it is," not "in the brain." The brain, serving a reconstructive wave passing through the holographic field is specific to a source within the field, now, by this process, an "image" of the source, that

FIGURE 3.3. The brain serves as the reconstructive wave, passing through the holographic field, specific to a source within the field, now by this process an *image* (here, of the cup).

is, the image is an *aspect* of the otherwise *nonimageable* holographic field (just as the information on a hologram is not imageable)—an *image* because it is *not* the "thing-in-itself," *not* the *totality* of the coffee cup as it exists with all its relations (real actions) within itself and with this field. This is the optical mechanism supporting direct perception.

Thus far, we have examined one of the aspects of the two-fold gap in the resonance model we noted in Chapter 2 ("Gibson and the Resonating Brain"), namely, the phenomenal aspect of the gap, or equivalently how the image of the external world is specified. But the temporal aspect of the gap has still to be explained, namely, the fact that the image is of *the past*, that the fly's wingbeats, seen in our perception, have long ago gone into the past as the universal field continues to transform over time.

THE CLASSIC METAPHYSIC OF SPACE AND TIME

Gibson's concept of time is extremely coordinate with Bergson's framework of time, formulated already in his *Time and Free Will* in 1889. This is little understood, even by Gibson's ecological school, but this view of time is what underlies the coherence of that vision of the brain resonating in symmetry with, and reflective of, the external event and the invariance structure describing both—dynamical brain and ongoing event [Robbins2023]. Gibson's statements on time are not all that prolific, but he certainly expressed a definite warning regarding importing physics concepts of space and time into psychology. Gibson notes that our implicit framework is that "space" is perceived, but "time" is remembered, and then warned, "But these abstractions borrowed from physics are not appropriate to psychology" ([Gibson1966], p. 276).

Although this statement seems almost innocuous, and has certainly been ignored, we will argue this is far from the case. This is in fact a strong warning, in reality a prohibition for psychology on relying on and employing physics concepts of space and time, that is, its "abstractions" re space and time—which is really to say the *metaphysic* of space and time upon which rests the mathematics of physics framework of explanation with its calculus. And when we say "psychology" here, we are equally addressing the theorists of AI, for if there is any pretension in AI to equaling (in the sense of being a theory for) human perception and intelligence, then AI is entirely linked to the science of psychology.

The chapter in which Gibson's statement is embedded, if not most of the entire work, is an analytical exercise in examining why these abstractions cannot hold. But Gibson was implicitly saying that psychology, and obviously ecological psychology, must work within a different metaphysic of space and time. It is only within this framework that his theory of perception (termed "direct perception") obtains true comprehensibility, and as well, the implied and allied notion of *direct memory*—with his seemingly strange declaration in this same book that taking the brain as a "'storehouse' of memory … is stultifying" ([Gibson1966], p. 277). Just preceding his warning re physics space and time "abstractions," Gibson had noted that the notion that memory applies to the "past," while sense perception applies to the "present," is "wholly introspective," for information does not exist exclusively in the present. One can view the traveling, razor's edge of the present as did William James ([James1890], Ch. 9, p. 15), the future and past extending on either side of this abstract point, but this is not the process of perception. For Gibson, "Physical events conform to the relation of before and after, not to the contrast of past and future" ([Gibson1966], p. 277).

This passage is of far more import than is understood; it is describing the inescapable – and unusable for psychology—consequences of the classic (spatial) metaphysic. This metaphysic, as Bergson [Bergson1896] described it, is an *abstract space*, a "principle of *infinite divisibility*." Beneath (or through) the concrete extensity of the physical world, we imagine a continuum of points/positions (Figure 3.4). A motion of an object, say a cup, from point A to B through the continuum is envisioned as a series of points—a trajectory—a line (i.e., again a *space*). As this space/continuum is infinitely divisible (and thus the line as well), we can imagine between each pair of points successively occupied (or passed thru), yet another line of points, also infinitely divisible, and between each pair of points on this line, yet another line ... Already, one can see, motion treated in this way—as a series of *immobilities*—is both an absurdity and an infinite regress.

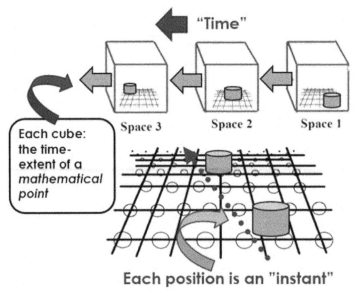

Each position is an "instant"

FIGURE 3.4. Successive positions of the moving cup across (or thru) the 3-D continuum of points/positions. Each point/position of the cup corresponds to instant of the all of Space.

It is this infinitely divisible space, Bergson argued, that is at the heart of all of Zeno's paradoxes: In a scenario (the 4th paradox) where two objects approach each other, both passing a third object that is stationary, Zeno, looking only at the *space* traversed, states that a "duration is the double of itself," or an arrow (3rd paradox), always at a static point of the continuum, "never moves," or (2nd paradox) Achilles, constantly dividing the intervening distance in half, thus into ever smaller intervals, "never catches the tortoise."

> At bottom, the illusion arises from this, that the movement, once effected, has laid along its course a motionless trajectory on which we can count as many immobilities [static points] as we will. From this we conclude that the movement, while being affected, lays at each instant beneath it a position with which it coincides. We do not see that the trajectory is created in one stroke, though a certain time is required for it; and that though we can divide at will the trajectory once created, we cannot divide its creation, which is an act in progress and not a thing. ([Bergson1907], p. 309)

And Gibson, contemplating the same problem, would do so in terms of a traveler looking ahead at the path to be traveled, looking behind at the path already traversed, with the point in between termed "here." Thus, "The *traveler is tempted to think of the linear path as the dimension of time,*" ([Gibson1975], p. 300, emphasis added). Emphasizing the confusion with space, the (linear) path traveled is the past, the future is the path ahead, and the spatial division point is taken to coincide with the instant "now."

Yes, it has been argued in the Achilles case that this is resolved by taking the limit of a converging series (e.g., Whitehead, [Whitehead1929]). But taking a limit is merely a *mathematical convention* allowing us to arbitrarily stop what is in reality an infinite operation or division. Even the wave equation, ψ, supposedly the basis of physical reality, harbors this mathematical sleight of hand. Intrinsic to it is Euler's identity, $e^{i\pi} + 1 = 0$, but this identity requires taking a limit to reach that "zero," to then begin the wave cycle again.

Besides this problem of the artificiality of taking a limit, Bergson noted that this mathematical treatment of Achilles is dealing only with divisible *lengths*, that is, with the *ends* of intervals, not the motion within an interval. In other words, it deals only with space, not time. But as Lynds [Lynds2003] noted, there is no interval, however, no matter how minute, that Achilles is not *passing through*. The mathematical framework is only a static backdrop to which Achilles has no actual physical relation.

Each point along the cup's trajectory is taken to correspond to an instant in time. Time is simply the 4th dimension of this abstract space. If that instant is taken as an instant of the all of Space (envisioned as a Cube, Figure 3.4), time is a series of such spaces or Cubes and given there is an (arbitrary) "end" to the infinite division, each Cube has the logical time-extent of a mathematical point (a *point* defined as having no beginning or end, thus no longer divisible). In effect, this is a frozen Cube, in fact a frozen universe; no further change is possible; there is no way to transition from Cube 1 to Cube 2. It was in contemplating this implication, Bergson noted, that to account for change, Descartes felt the intervention of God was necessary (presumably *creating* Cube after Cube). Simultaneously, the frozen Cube—again, a mathematical point in time-extent—is the one condition in which a fixed, determinate value for Achilles' motion is possible, else, again, the value is always *uncertain*; he is always passing through the interval, no matter how small. But the price of obtaining this determinant value is, again, no more change. As Lynds [Lynds2003] noted, echoing Bergson, to enable a changing universe, this is an intrinsic tradeoff—uncertainty for constant change.

In this instantaneity, as such, each Cube is utterly *homogeneous*. This (i.e., the metaphysic) is already the core of the "hard problem" [Chalmers96], for no *qualities* can arise in such a space or in a series of such spaces, where each space instantly vanishes as the next (the "present") arrives, certainly no qualities of *motion* such as Valerie Hardcastle pictured in her description of qualia with those musicians concentrating, patrons shifting and the curtains gently and ever-so-slightly waving.

In this classical framework, there is ever only one Cube—the present—the previous Cubes each having successively fallen into the past, the "past" being our symbol of *nonexistence* in this metaphysic. The brain is equally just a "subcube" within these Cubes of the universe—sub-cube after sub-cube. Any event—stirring coffee—subcube after subcube. There is no actual *time extent* of any event *nor of the brain taken as an event*. It is this framework that Gibson is implicitly reacting to. Thus, "Resonance to information, that is contact with the environment, has nothing to do with the present" ([Gibson66], p. 276), that is, nothing to do with this scheme of "present instants" and the artificial *ideal limit* the present instant represents within the metaphysic.

Gibson goes on, making the problem more explicit, noting, with James, that the traveling moment of present time is certainly not a razor's edge, and that the question of when perception leaves off and memory begins is yet unanswerable. He notes that Boring [Boring1942] described the (unsuccessful) efforts to get around this question in his chapter on the perception of time. The simple fact is that, "… animals and men perceive motions, events, episodes, and whole sequences" ([Gibson1966], p. 276), and perceiving is not focused down to the present item in a temporal series. Then he alludes to the problem of short-term or immediate-term memory models explaining time-extended perception—the coffee stirring—for these models require a basic assumption, namely, that a succession of items can be grasped only if the earlier ones are held over to be combined with later ones in a single composite. "This," Gibson notes, "can be pushed into absurdity" ([Gibson1966], p. 276). "Pushed into absurdity" because, as we earlier noted, subcube after instantaneous subcube of the coffee stirring (or its "features") must be stored in a static space (supposedly in the brain), like a series of 3-D snapshots laid out on a desktop—a vast, static structure. But now, how does one reintroduce the motion? As we saw Turvey asking [Turvey1977a], via some "internal scanner"? Another, already noted (Chapter 2), is invoking "the continuity of neural processes," for example, "… *the extended temporal duration of activity necessary* for consciousness" ([Taylor2002], p. 11, emphasis added).

Now we can point out that this is conveniently *ignoring the logic of the metaphysic* in which the theorist is working wherein the brain itself can only have the time-extent of a mathematical point (and thus, equally, *those neural processes*). This—the ever-disappearing present—is precisely the source of the seeking and theorizing as to how experience is stored in the brain in the first place, for as "matter" is defined as that being always "present" and the brain is matter, thus deemed always "present," the brain (yes, incoherently) is deemed the only place for safe storage of the nonexistent past.

THE TEMPORAL METAPHYSIC

So, how does the brain *specify* (as a reconstructive wave) time-extended, dynamically transforming events—the coffee being stirred, spoon circling, surface swirling? As noted, Gibson is implicitly invoking, or better—requiring—a different metaphysic. Bergson had already laid this out.

Below homogeneous [abstract] time, which is the [spatial] symbol of true duration, a close psychological analysis distinguishes a duration whose heterogeneous moments permeate one another; below the numerical multiplicity of conscious states, a self in which succeeding each other means melting into one another and forming an organic whole. ([Bergson1889], p. 128)

He would also compare this flow to a melody, where each "note" (read "instant") permeates the next, where the state (or quality) of each note reflects the entire preceding series, and where these comprise an organic continuity.

Motion, he argued, must be treated as *indivisible*. When, per Zeno, Achilles successively halves the distance to the tortoise, it is his track in space, the infinitely divisible line, of which we think. Rather, Achilles' motion (the *process*) is indivisible; he moves with indivisible steps, he most certainly catches the tortoise. Per Zeno, the arrow, always being coincident with a static point on this infinitely divisible line, "never moves." But the arrow in fact moves in an indivisible motion. When a "duration is the double" of itself, as in *all* these paradoxes, it is the abstract space that Zeno is considering.

The abstract space of the classic metaphysic with its mathematical treatment erases *real*, concrete motion. The cup can move across the continuum (or coordinate system), or the continuum move beneath the cup. Motion now becomes immobility dependent purely on perspective. All real, concrete motion of the matter-field is now lost. But, Bergson argued, there must be *real motion*. The universe, the entire matter-field, must dynamically change and evolve over time. Trees grow. Roses bloom. People get older. Mountain ranges appear.

Stars shrivel and die. He would insist then, already acknowledging the only partial validity of a relativistic point of view:

> Though we are free to attribute rest or motion to any material point taken by itself, it is nonetheless true that the aspect of the material universe changes, that the internal configuration of every real system varies, and that here we have no longer the choice between mobility and rest. Movement, whatever its inner nature, becomes an indisputable reality. We may not be able to say what parts of the whole are in motion, motion there is in the whole, nonetheless. ([Bergson1896], p. 191)

We must, he argued, view the entire matter-field as a global motion over time. We could visualize the whole changing, he argued, "as though it were a kaleidoscope." We want to ask if individual object X is at rest, while individual object Y is in motion. But both "objects" are simply arbitrary partitions, phases in this globally transforming field. As such, the "motions" of "objects" are seen as *changes or transferences of state*—rippling waves—within the dynamic, indivisible motion of the whole.[1]

So, again the question: how can time-extended events—events extending into the "past" —be specified? From the perspective described this property of the dynamically transforming holographic matter-field itself and of its melodic, indivisible transformation, where every "instant" (or "note) permeates the next, defines a "primary memory" (appropriating the term from James, [James1890]). This primary memory underlies the motion of the rotating cube, the swirling coffee and stirring spoon, the motion or flow in the neurons of the brain. The motion of the field, of which the rotating cube is just a phase or transference of state, does not consist of discrete instants ("presents") that fall away, one by one, into the past, into nonexistence. For this reason, the brain, as a resonant (reconstructive)

---

[1] The indivisibility of the time-transformation of the matter-field will seem at odds with the concept of "quantum jumps." We suggest Schrödinger's powerful destruction of the "jump" concept [Shröd1952a], [Shröd1952b]; for a summary [Robbins2021]. And we should note the 2023 Physics Nobel Prize for techniques allowing attosecond scale detection and subsequent experiments [Quanta2023] which overthrow the "instantaneous transitions" (jumps) concept. These developments certainly appear as confirmation of Schrödinger.

wave embedded in this transformation (and equally time-extended) in Gibson's metaphor, is able to specify these transformations of the matter-field, even though from the standpoint of the classic metaphysic, they are now long in the "past." It is how we can be dealing with *successive order*, not with a series of "presents" each constantly moving into the past. The brain, as a reconstructive wave, can specify "rotating" cubes, ever so gently waving curtains or the "singing" notes of violins. Yes, *qualia of motion*. To answer Gibson's question, there is no "dividing line" demarking when "perception leaves off and memory begins." We are always viewing the "past." Perception, as Bergson noted, is always already a memory.

In such a flow, we should note, each instant interpenetrating, reflective of the preceding series, there is the foundation for qualities that only become so by *building in time*, for example, the "mellowness" of a violin, of a wine, or of a being. And nothing, due to this building, *ever precisely repeats*. Even in the classic case— the same force applied by the same cue to the same billiard ball with the same directional vector as the resultant over and over— the cue is never actually the same, the ball is never the same from time 1 to time 2—only *practically* the same. This is why the classic metaphysic, in describing this qualitative transformation of the field, employing spatial conventions such as Euler's identity—a cycle always coming back to zero, that is, to the *precisely same state*—is *an ideal limit*, and can ever only *approximately* capture the real nature of the transforming field.

THE SCALE OF TIME

We have already alluded to another temporal aspect of this specification that should be registered, namely, that the specification is to a particular *scale of time*. The matter-field—the *environment* of ecological psychology—is not limited to our normal scale of "buzzing" flies, butterflies flapping their wings, grass fields barely perceptibly waving. This normal experience of ours is a particular scale of time. A Minkowski space-time diagram with its light cones, as we noted, can legitimately be depicting a manifold of nothing more than electrons whirling. Something must impose or "specify" a scale of time upon the manifold, and this is the resonant wave that is the brain. This time scale-specification, it is reasonable to assume, is based in the underlying chemical velocities of the neural processes, where chemical velocity has its standard expression in the Arrhenius equation. Such a velocity can be changed, as simply as by raising the temperature, or by introducing some form of catalyst. How LSD, for example, might work as such a catalyst, biochemically, in changing the normal, specified scale of time, will be discussed later. The point is, one can imagine steadily *increasing* the brain's underlying chemical velocities, and in such a scenario our normally "buzzing" fly transitions to a heron-like fly slowly flapping his wings, and with a greater increase of chemical velocity, to a motionless fly whose vibrating crystalline structure can begin to be perceived, and on.

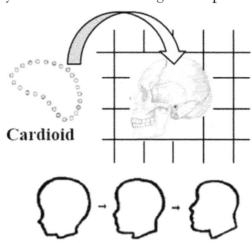

FIGURE 3.5. Skull profiles aging under the strain transformation. A cardioid, fitted to the skull, is stretched in all directions like a rubber sheet. (Adapted from Pittenger and Shaw, 1975.)

The buzzing fly, the heron-like fly, the motionless, crystalline vibrating fly are qualia—qualities of form and motion [Robbins2004] [Robbins2013]. This—and the intrinsic role of time itself—has been lost on the theorists of the hard problem, but this form of "qualia" is a natural consequence of the brain as resonantly specifying the environment—at a scale of time. In essence, in the thought experiment above we have done the analog of changing the relativistic "space-time partition." In special relativity, it is invariance laws only that hold across these partitions, for example, the law $d = vt$ in the stationary system becomes $d' = vt'$ in the moving system. The same invariance laws that specify the buzzing fly also specify the heron-like fly, or concomitantly, the coffee being stirred in the buzzing fly partition as well as the slower coffee swirling and stirring in the (much slower) heron-fly partition, or in Shaw's original context, the same law holds that is specifying the aging of a facial profile, namely, a strain transformation on a cardioid (Figure 3.5, [Pittenger1975]), whether for a very slow event (as is normal) or a much faster (rapidly aging) event.

When taken in conjunction with the brain's necessary role in specifying the environment at a scale of time, the virtual action principle (and equally so, or should have been so, for the "affordance" concept) indicates that as the specified time-scale changes, the possibility of action is equally changing, and this must be so for perception to be *ecologically valid* at all scales. If the "buzzing" fly is an index for the kind of action possible, say, to move the hand-arm quickly to *grab* the fly, then a heron-like fly or even better, a nearly motionless fly, is a specification of a different form of possible action, for example, reaching out slowly and leisurely grasping the fly by the wingtip. Such an implication, we will argue below, should be ultimately testable.

DIRECT SPECIFICATION AND THE PROBLEM OF ILLUSION

As noted previously, the specification is always to the past of the field. The fly's wingbeats being specified have long gone into the "past," but the indivisible motion of the field with its intrinsic, primary memory, supports this past-specification. The fly's motion with its wingbeats does not consist of a series of instants, each of which immediately enters the non-existence of the past as the next instant—the "present" instant—arrives. The image, too, is right where it says it is—in the field. It *is* the field—the (specified) past of the field—at a specific scale of time. Similarly, even one second of the strobed, rotating cube is, in reality, a vast series of (interpenetrating) states within the external, ever-transforming holographic field, with each "state" in this series containing a slightly different orientation of the cube. Under the out-of-phase strobe, the optimal specification is to a superposition of "states" in this series (at a scale of time) wherein the rigid edges and vertices are lost—now being the wobbly, plastically deforming not-a-cube. Again, the brain is not projecting or generating an image.

The plastically deforming not-a-cube would be considered an *illusion*, for we "know" there is actually a rigid, rotating cube involved. But even illusions, the great redoubt of indirect realism—the ubiquitous conception that the brain must be (somehow) *generating* the image—are but optimal specifications of events (sources) within the field, where the "image" is precisely where it says it is—in the external field. Even the lack of processing during the rapid eye-movement of a saccade from point to point on a scene, argued by indirect realists to prove that the brain must be generating an image to maintain the illusion of a continuous world, can be simply countered by noting that the reconstructive wave supported by the brain need not cease during a saccade.

Indirect realism's argument from illusion, to conclude that the brain is always generating the image of the world—*all* experience in essence being a vast hallucination—essentially is assuming a God's eye view. In Figure 3.2, we "know" that the reconstructive wave, modulated to f_1 should specify the cube, and modulated to f_2, the cup. A less coherent reconstructive wave, modulated to a non-coherent (nonunique or composite) frequency of f_1 and f_2, specifies a fuzzed superposition of the two scenes – fuzzy cup fused upon the cube. Our omniscient observer status allows us to say this fuzzed scene is an "illusion." The brain, as a reconstructive wave, intrinsically embedded within the transforming holographic field, has no such privilege. There can be no God's eye view selection.

Currently, in perception and consciousness, there are a couple favorite themes:

1. Color does not exist in the material world. Color is only "in the mind," or in some "mental space." In other words, all color is in essence a brain-generated illusion. There are only non-colored particles—homogeneous little billiard balls.

2. There is an "inverse problem," and it is all solved by a Bayesian approach. (The retinal information is so ambiguous—so vague in structure—the external world must be guessed via probabilistic processes.) In other words, all perception is just a brain-generated guess-construction.

One can see, again, that this reflects the drive to indirect realism. This is in support of a Kantian "constructivist" conception, furthered by Helmholtz (for whom all perception is an *inference*), and refined further in *Predictive Processing* by Karl Friston (e.g., Brown and Friston, 2015 [Brown2015]) and others where the inferences are instantiated via neural networks as Bayesian inference processes. In this, given Bayesian probability "priors" which capture prior knowledge about the environment (e.g., "light almost always comes from above"), likelihoods can be calculated and thus posterior probabilities which guide the selection of a percept—"[Given the light is from above] we are seeing a *convex half-sphere* on the ground," (as opposed to a concave hole, lighted from below).

Purves and Lotto [Purves2010] took a large wrecking ball to this program. How, one can ask, is the space of priors constrained? If I am about to perceive a cup of coffee, there are an enormous number of arrangements and looks for coffee cups—which of these is used as the prior? Or, how do we know there are only domes and holes in an environment of mountains, forests and lakes? The priors, Purves and Lotto argued, come via the information coming from *the bottom*, that is, from the environment! It comes from pure *statistical exposure*. Vision sees half-spheres because it is exposed to them, not because it is making inferences.

This brings us back for a moment to Gibson who had long argued that the information for perception is far richer than presumed by the inverse problem. "Illusions" are a function of an artificial, information-deprived arrangement where natural information normally in the environment has been destroyed. Implied in this already was in essence a statistical argument: There will be invariants in natural scenes that are statistical regularities as in the Poggendorf illusion or even the Ponzo illusion (Figure 3.6). In a natural environment, the stimulus information is highly constraining, limiting things to *one* "hypothesis."

Natural Scene Statistics researchers later initiated the look at statistical properties in the external environment. The findings for the specification of color are of critical interest here. For example, for two adjacent inner squares of color in different surrounds (Figure 3.7), the right square appears darker. Why? Because statistically, given the lighter surround, such a high-luminance surround co-occurs more frequently with high-luminance targets than with low-luminance targets. This means that the stimuli are more likely caused

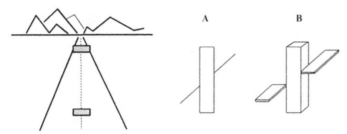

FIGURE 3.6. The Poggendorf Illusion (Right). The diagonal line of A is actually perfectly straight through the block. Vision presumes the 3-D configuration of B. Left: The Ponzo illusion: The two "blocks" are actually exactly the same width. Vision presumes they are placed on the ubiquitous texture density gradient. Then indeed the far block would be larger.

by the scene in which there is a lot of light over the right panel but less over the left. Again, there is no appeal to inference here, but rather, to the statistically usual causes of a scene. If the stimulus is normally produced by things of a different brightness, we're likely to see them as such. The inverse problem is explained by appeal to how the world is.

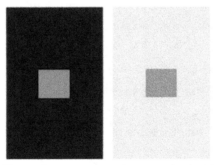

FIGURE 3.7. The two inner squares are the same color, but the right one is perceived as darker. (Adapted from Purves and Lotto, 2010.)

This simple example of the squares is simply the start of a long series of demonstrations by Purves and Lotto that all perceived color depends on this statistical frequency of conditions in which things are usually seen. A square grey patch on a cube becomes yellow or blue depending on the lighting setup. This yellow or blue when the square is actually grey – this perhaps is the quintessence of what drives theorists into (at least implicit) idealism, and especially, the near universal opinion: The brain creates color; color is "in the mind."

We end in a mystery never to be solved. Why? Driving this: the classic metaphysic of space and time. In this framework, all quality has been stripped from the material field. Each "Cube of Space" of Figure 3.4 is but a mathematical point in time-extent. As such, as noted, no qualities are possible. (Thus, no colored vibrations.) But, given the classic metaphysic, *the brain is equally—integrally— embedded and described in this classic (quality-less) conceptual framework.* As such, the brain is helpless, logically precluded from *creating* ANY quality, whether qualities of motion and sound: gently waving curtains, roses opening, mellow violins; qualities of timescale: buzzing flies, heron-like flies. We either end in this complete mystery with color residing in some vague "space," for example, a mental space or perceptual space, that is, anything to escape the dead, quality-less space of the metaphysic, or we accept that color is a property of the universal field and that the brain is not magically creating it—*specifying* (for the sake of action), employing memory/ experience—yes. Creating, no.

The problems are of a piece: color, dynamic form, motions. The holographic field is intrinsically qualitative: Spread out the "buzzing" fly in time (within one objective second as measured by the second-hand of a clock):

- buzzing—wingbeats (200/sec) = a blur

- flapping—wings moving slowly like a heron

- motionless—starting to see shimmering, crystalline oscillations of the fly's body

- motionless—an ensemble of swirling electrons

Spread out the color red (over 1 sec) in objective time:

- One second of red light is 400 billion wave oscillations.

- "Spread out" —given our ability to discriminate events in time, it would require 25,000 years to count each wave.

In other words, in "spreading out" the light (just as the fly), in moving toward each individual wave, we are moving nearer and nearer to the abstract space—the homogeneous Cubes taken at the smallest conceivable instant/extent of time. But this is an *ideal limit*—NEVER reached. This is Bergson on the subject:

> May we not conceive, for instance, that the irreducibility of two perceived colors is due mainly to the narrow duration into which are contracted the billions of vibrations which

they execute in one of our moments? If we could stretch out this duration, that is to say, live it a slower rhythm, should we not, as the rhythm slowed down, see these colors pale and lengthen into successive impressions, still colored no doubt, but nearer and nearer to coincidence with pure vibrations? ([Bergson1896], p. 268)

Color is also an optimal specification, just as the buzzing fly—at a scale of time—of vibrational properties of a qualitative field. It is how the brain as a reconstructive wave specifies the information in the field. Color is not "created" by the brain; it is not an illusion residing in some "mental space."

A CONCRETE RESONANCE

This, in brief, is a framework in which Gibson's "specific to" gains a concrete coherence. Yes, the detailed description/science of how the brain in effect forms a modulated reconstructive wave certainly awaits, along with the physics of the holographic field (Bergson is *not* "the holographic principle" of current physics). This eventual description of the brain's dynamics must intrinsically incorporate the dynamic invariance structure of the external event, for the argument here, one in effect Gibson already has made, is that *it is the invariance structure of the event* that is driving, modulating the brain as this dynamic, specifying, modulated reconstructive wave. We need to remember that in the coffee stirring case, this list of invariants included radial flow fields, an adiabatic invariant, an inertial tensor, a tau ratio, a constancy as defined over the texture density gradient relative to cup height, acoustical invariants, and more.

Certainly, all this is related in the brain to the action systems, necessarily so for perception being the display of possible (virtual) action. Again, a description of the neuro-dynamics of the brain that does not show how the dynamic invariance structure is reflected in its processes, in its resonant wave, must be considered off the track [Robbins, 2014]. We saw Raja [Raja2018] reflecting this in his statement that "… resonance is what is going on inside the organism, especially in the CNS, with regard to what is going on at the ecological scale (p. 33)," and the example of the invariants specific to the baseball's arc through the air (and simultaneously to the outfielder's catching the ball) being reflected in the neural processes of the brain. However, as noted, there is the distinct possibility that one can alter the biochemical level of the CNS, that we can raise the resonant frequency, so to speak, of the CNS, and now what is going on in the CNS is relevant no longer "to what is going on at the ecological scale," at least to the *normal* ecological scale of humans, although perhaps it is now what is normal to frogs with their higher metabolic rate—a function perhaps of the ratio of body mass to oxygen consumption [Fischer1966]. As opposed to "buzzing" flies, the flies now specified are nearly motionless, stable flies (easily flicked out of the air by a frog-tongue in *its* normal scale). We note this to again make the point that *what the brain is specific to* is equally critical; the invariants hold across all scales to time.

Raja rejects the current, standard, cognitive science-move or transition from a description based in the concrete physics and dynamics of the brain to a level of description that is purely computational. He moves to a description of the CNS and the invariants of the environmental event based in DST, an approach which, as noted, given the right parameters, nicely demonstrates the symmetric reflection of the structures. We would simply note that ultimately, this resonance (and Bergson's reconstructive wave) must be taken *far more concretely* than DST can capture. Yes, one can describe an AC-motor in terms of attractors and bifurcations, but this is ultimately unhelpful if building an AC motor, for now the actual forces and materials and configuration thereof become critical.[2] When describing the reconstructive wave in Bergson's framework (and which we are arguing is in effect Gibson's) — the forces and materials supporting it—this concrete level will be required, for the brain is supporting or creating a very concrete wave "passing through" a very concrete holographic field. When we enter the biochemical level relative to perception, as we shall see shortly, this concreteness and its complexity becomes especially evident.

---

[2] A fact Charles Steinmetz knew well when trying to formalize principles required for the construction of N. Tesla's AC motor (*The Theory and Calculation of Alternating Current Phenomena*, 1900), up to that point nearly nonreplicable (they were burning up) because of the precise nature of the metals involved–a knowledge, for a period of time, apparently peculiar to Tesla.

AFFORDANCES: THE IMAGE AS VIRTUAL ACTION

The continuous modulation of the brain (as a wave) is driven by the invariance structure of the transforming external events, for example, the velocity flows defined over the sides of the cube as it is rotating conjoined with its recurring symmetry period. Due to the continuous motion of the field, this information is always inherently uncertain—we have always an optimal specification of some subset or aspect of the past motion of the external, holographic field. There is no "veridical" or God's eye view selection. The brain, as a reconstructive wave, is selecting information from the ever transforming, holographic matter-field, where the principle of selection is based on information (invariants) relatable to the body's action systems.

This, it can be argued, is a significance of the intimate feedback to and from the brain's motor areas to the visual areas as noted by Churchland et al. [Churchland1994], that is, the modulation of the visual areas by the motor areas. It is consonant with the finding that perceptions reflect the body's biologic action capabilities [Viviani1990], [Viviani1992]. It is also likely the significance of the fact, insufficiently noted with the exception of Weiskrantz [Weiskrantz1997], that upon simply severing the tracks from the motor areas to the visual areas, the monkeys undergoing the procedure went blind [Nakamura1980], [Nakamura1982]. With no connection to action, there can be no selection, no virtual action, and no vision.

There is also a deeper implication here in this principle of virtual action—ultimately testable. We have already alluded to the possibility of introducing a catalyst into the biochemical makeup of the body/brain dynamics, with effects on the perceptual "space-time partition." In the possibility of the fly becoming a heron-like fly, slowly flapping its wings, we already previewed the relativistic aspect of this principle. The fly is always presented to us at a certain scale of time. Let us complete the implications, for the time-scaling of the external image is not a merely subjective phenomenon—it is objective and has objective consequences realizable in action.

Let us picture a cat viewing a mouse traveling across the cat's visual field (Figure 3.8). As we have seen, there is a complex,

Gibsonian structure projected upon the cat's retina. There is a texture density gradient over the surface between mouse and cat. The size constancy of the mouse as it moves is specified by the constant ratio of texture units it occludes, while the tau value is specifying the impending time to contact with the mouse, with its critical role in controlling action. All this and much more is implicitly defined in the brain's "resonating" visual and motor areas. How is the information related to action we must first ask?

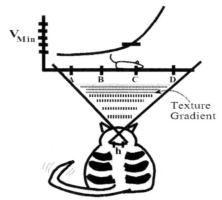

FIGURE 3.8. Hypothetical function describing the minimum velocity required for the cat to intercept the mouse at D.

Turvey [Turvey1977b], an ecological theorist of perception and motor action, described a "mass-spring" model of the action systems. For example, reaching an arm out for the fly is conceived as the release of an oscillatory spring with a weight at one end. "Stiffness" and "damping" parameters specify the endpoint and velocity of such a spring system. These constitute "tuning" parameters for the action systems of the body. The tuning parameters "bias" the hand-arm to release at certain velocity and stop, just as a coiled spring.

The track (ABCD) along which the mouse runs projects upon the cat's retina (the little "h" in Figure 3.8). For computing the velocity of the mouse, we need: velocity = distance/time. Here the distance (d) being traversed on the retina is equivalent to h (the track/line projected on the retina). But if we wish to use h for distance in our equation to obtain the velocity of the mouse, we have an ambiguity. Similar to the problem with Berkeley's line ABCD in Figure 2.4 (Chapter 2), we can move this horizontal track, along which the mouse runs, even closer to the cat, yet the horizontal projection (h)

on the cat's retina is exactly the same. Any number of such mice/ tracks at various distances project similarly to h; h is always the same. To solve for the velocity of the mouse, the needed muscle-spring parameters must be realized directly in the cat's muscular structures via properties of the optic array, for example, the texture density gradient across which the mouse moves and the quantity of texture units he occludes.

At our normal scale of time, we can envision a function or curve (in Figure 3.8) relating the minimum velocity of leap (V_{min}) required for the cat to leap and intercept the mouse at D as the mouse moves along his path. The closer the mouse gets to D, the faster the cat must leap, until the minimum velocity is so high, it becomes impossible. But to compute the velocity of the mouse, the body needs one other critical thing. It needs a standard of time.

A physicist requires some time-standard to measure velocity. "Objective" time, it should be understood, is simply the noting of spatial simultaneities. Each time the earth returns to, and coincides with, a point in its orbit, a "year" has passed. This is the noting of spatial simultaneity. The physicist then could use a single rotation of a nearby rotating disk to define a "second." Each time a mark on the rotating disk passes a certain point/mark on the table, a second is declared to have passed. But if an evil lab assistant were to surreptitiously double the rotation rate of this disk, the physicist's measures of some object's velocity would be halved, for example, from 2 ft/sec. to 1 ft/sec. The body though must use an internal reference system. Similar to the rotational velocity of the disk, this internal system must be an internal chemical velocity of the body. Such a system is equally subject to such "surreptitious" changes.

We noted that these velocities can be changed by introducing a catalyst (or catalysts)—an operation that can be termed, in shorthand, modulating the body's "energy state." If we raise the energy state, the function specifying the value of V_{min} for the cat must change. There is a new (lower) V_{min} or minimum velocity of leap defined along every point of the object's trajectory, and therefore the object, if perception is to display our possibility of action with ecological validity, must appear to be moving more slowly. The mouse is perceived as moving more slowly precisely because it reflects the new possibility of action. If this were not the case, we would be subject to strange anomalies. The cat would leap leisurely; the mouse would long be gone. The cat's perception would not be ecologically valid; it would not specify the appropriate action available.

If the fly is now doing his heron-like imitation, flapping its wings slowly, the perception is a specification of the action now available, for example, in reaching and grasping the fly perhaps by the wingtip. It is equally a reflection, we might say, of the new tuning parameters for our "mass-spring" hand-arm. In the case of cube spinning so rapidly it is a cylinder with fuzzy edges (in this circular form, now a figure of infinite symmetry), if by raising the energy state sufficiently we gradually cause a perception of a spinning cube with serrated edges (perhaps 16 edges, then 12 edges) and then the four-sided cube in slow rotation, it is now a new specification of the possibility of action, for example, of how the hand might be modulated to grasp edges and corners rather than a smooth cylinder.

Again, the requirement comes to the fore for invariance laws holding across these space-time partitions of perception, in the rotating cube's case, the invariant figure of 4n-fold symmetry in any partition. The hand's modulation is guided by the same invariance information across partitions. Again, we can consider the aging of the human facial profile – a very slow event in our normal partition (Figure 3.5). As Pittenger and Shaw showed, the aging transformation is a strain transformation upon a cardioid. We could imagine this aging event as greatly sped up, perhaps as though for a Mr. C who dwells in a much lower energy state where lifetimes pass very quickly. For Mr. C, watching another human, the head transforms very quickly. Yet, for Mr. C, the aging event is specified by exactly the same invariance law, and his action systems will use the same information should he reach out to grasp the rapidly transforming head.

That perception is the display of possible (virtual) action, that is, of veridical action across scales of time, that, as opposed to the "buzzing" fly of our normal scale of time, a perceived heron-like fly specifies the possibility of grasping it by the wing-tip—this is a testable implication of the principle of perception as virtual action. The theory described here naturally generates it. We can think of no others that do, although in retrospect the implication will seem obvious. For the computer model of mind, it would merely be an *ad hoc* afterthought or theoretic "epicycle," not that the computer model can account for the image of the external world in the first place, let alone its scale of time. This is not to say that our current understanding of the "catalysts" at work is anywhere close or will be easy, but the prediction is testable in principle, nonetheless.

REAL VIRTUALITY AND AI

The deep implications of virtual action for AI should be understood. It is a standard view of AI, robotics, and cognitive science that there is what is termed a "perception-action cycle" (Figure 3.9). MIT Robotics professor, Illah Reza Nourbakhsh describes it in terms of intelligence being meaningfully connected to the environment, and internal decision-making skills to consider our circumstances and then take action. Thus, per him, "The environmental connection is two-way, and we term the inputs as perception and the outputs back to the world as action" ([Nourbakhsh2013], p. xvi). Cognition, then, is the internal decision process that transforms our senses about the world into action, an action which then changes the world, ready to be sensed again.

Thus, we perceive, cognize and act, and cycle again. There is nothing exactly wrong with this—we indeed do perceive, then contemplate, then act—it just captures nothing of virtual action; the standard formulation is oblivious to this reality. The problem has already begun for AI in Nourbakhsh's statement on "being meaningfully connected to the environment." This is affected, for the human, by the fact that the image of the external world is simultaneously—very meaningful to the organism—virtual action. The image is both a display of the past and an orientation to the future.

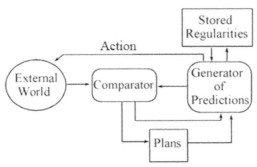

FIGURE 3.9. A variant of the perception-action cycle.

We can put this in a context for robotics that makes the implication concrete. Place the problem in the context of the Turing Test, the nice test described by Alan Turing that he thought might be useful in determining when machine intelligence had arrived at the point where it is at least indistinguishable from the human. Turing envisioned a human asking questions of a computer behind a screen, with the computer printing out answers. But we can extend the test a bit. Thus, an extended question might be, "Mr. Robot (or Mr. Human—we're not sure of course), please ingest this small pill or tablet I'm pushing under the curtain [containing a catalyst or set of catalysts of a certain strength]. Now please describe this little fly I am sending by you. Is it "buzzing," or flapping his wings like a heron, or an ensemble of whirling particles?" Mr. Robot of course would be expected to give an answer within the understood human parameters of how the fly should appear.

VIRTUAL ACTION—THE DIFFICULT DETAILS

Bergson's selection principle—that a subset from the mass of information in the holographic field is selected via its relevance to the body's action—is the bedrock, the point where one must start in this model of vision. There must be a principle of selection. The implementation of this high-level principle in the brain, therefore its precise meaning, is something that requires more uncovering. Alva Noë [Noë2004], in his concept of vision as a function of "sensori-motor contingencies," seems extremely close to Bergson, if not in fact explicating virtual action more deeply. For Noë, a chair would have a set of sensori-motor actions associated with it–how we can sit on it, how it transforms as we move around it, or as we move it, and on. A ketchup bottle would have an entirely different set of actions, transforming differently as we rotate it, squeeze it, and so on. It is this different set of actions, argued Noë, that underlie how the bottle looks versus how the chair looks. For Noë, there is no internal representation, no model of the chair nor of the bottle in the brain. The image of each arises because the body is in a dynamic action/ feedback loop with the external field in which the bottle and chair reside, while the action contingencies set, for each, the character of the visual experience.

We can see, given our discussion of Bergson and Gibson, that Noë lacks the concept of the holographic field and the brain as a reconstructive wave within it–elements required to give any coherence to the notion that an image is accounted for by simply having objects in a relation to possible actions. Missing too is any consideration of the nature of dynamic form, as was discussed earlier, which indicates that transformations, laws, uncertainties and time are also part of the story. He is also simply assuming qualia via this dynamic loop, that is, he is taking an unacknowledged ride on Bergson's temporal metaphysic to account for qualia or the qualitative nature of the image. He has in his theory, to our knowledge, no explicit overthrow of the classic metaphysic and its model of (abstract) time. But Noë's concept of sensori-motor contingencies appears to convey the need, when we look at the chair, for this set of potential actions to play, in some mini, reduced, or virtual form, in the brain.

Andy Clark [Clark2008], in his attentive examination of Noë, called attention to the findings of Milner and Goodale [Milner1998] on the role of the dorsal and ventral streams in the brain, and the phenomenal effects of damage to each stream. Each "stream" is sourced originally from the optic nerve, and comprises a pathway of areas in the brain, one (the dorsal) branching from the primary visual area (V1) and projecting to the posterior parietal (PP) cortex, the other, the ventral, branching from V1 and projecting eventually to the inferior temporal (IT) cortex. The dorsal stream appears tuned with information to carry out the fine *details* of action, as when we pick up a letter, rotate it, and put it into a mailbox slot at the post office. The ventral appears to carry representations of objects oriented to higher level operations–"comparing, sorting, classifying" or the selection of possible actions. This latter, the selection of possible actions as information in this stream, Clark termed a *sensori-motor summarization*, that is, a high-level or virtual summary of a set of actions.

Phenomenonally, Milner and Goodale discussed a woman (identified as "DF") with damage to areas of the ventral stream, who reported that she could not "see" the orientation of a slot (like our mailbox slot) yet could rotate and insert a letter precisely into the slot. This indicates that the dorsal stream (undamaged in her case) carries information sufficient for effecting precise action, yet without connection to conscious vision, and in fact has nothing to do with conscious vision. On the flip side, folks with optic ataxia, a condition which results from damage to areas of the dorsal stream, would be able to see the slot and its orientation, but be unable to rotate and insert the envelope, that is, conscious visual awareness, but without the capability for finely attuned action. For Milner and Goodale (and Clark), this means that it is the ventral stream (and its representations of objects) that is responsible for conscious vision, and of course, as Clark noted, this called into question Noë's notion of a "playing-out" of sensori-motor contingencies as the undergirding for conscious vision.

We can leave it to Milner and Goodale and to Clark to explain how the "representations" in the ventral stream–encoded in their chemical flows–mysteriously become the image of the external world. It is important to note that in Clark's notion of "sensori-motor summarizations" in this ventral stream we are still very close to Bergson's virtual action. The curious implication is, if this profound dichotomy of the functions of the two streams indeed stands, that in regards to action in varying scales of time, if one were to increase the velocity of processes in the ventral stream but not in the dorsal, that is, not introduce a *global* change in process velocity, one would see the heron-like fly, but could only act as though it were the buzzing fly; one could not grasp the fly by the wing. But this sharp dichotomy of functions is not truly settled. Milner and Goodale allow that there must be connections between the streams, while Gallese [Gallese2007], on the basis of his findings (which also correlate phenomenal evidence with the locations of neural damage), argues for the existence of a ventral-dorsal stream. This pathway, part ventral, part dorsal (therefore involving action), Gallese sees responsible for the organization of actions directed toward objects (as in rotating the envelope toward the slot), but also for space and action perception and conscious visual awareness. For example, a lesion on one side of the brain (a unilateral lesion) of the ventral premotor cortex of the monkey, including area F4, (i.e., part of the ventral-dorsal stream and which controls the perception of space close to the person or our body), produces both motor deficits and perceptual deficits. Perceptually, a piece of food moved around the monkey's mouth, in the visual space opposite the side of the lesion, does not elicit any behavioral reaction. Similarly, when the monkey is fixating on a centrally placed stimulus, the introduction of food in the space opposite the side of the lesion is ignored. In contrast, stimuli presented outside the animal's reach (in far extra-personal space) are immediately detected. There is a similar pattern of phenomena when there is a lesion to areas responsible for vision in far extra-personal space, that is, space beyond our near-the-body (or peri-personal) space.

Gallese quoted the philosopher, Merleau-Ponty, who noted (echoing Bergson) that space is, "… not a sort of ether in which all things float …The points in space mark, in our vicinity, the varying range of our aims and our gestures" ([Merleau1962], p. 242). In other words, the abstract space of the classic metaphysic is meaningless to the body and perception. For Gallese, why is action important in spatial awareness? For him, action embodied simulation integrates multiple sensory modalities within the F4-VIP neural circuit. Since vision, sound and action form an integrated system, the sight of an object (or sound emitted) at a given location automatically triggers a "plan" for a specific action directed toward that location, and that plan to act is a simulated potential action.

There is clearly much to be dug out regarding the implementation of virtual action and/or Clark's sensori-motor summarizations. We agree with Clark that the concept of constantly running mini-simulations of actions as our eyes scan the table with its cups and plates and silverware (and at what scale of time? Every 1/10th second?) becomes incoherent. But there is a possibly illuminating question as to how the summarizations are initially formed. Clark's "summarizations" are not far from what is known as action "syntagms." For example, on seeing a spoon, we automatically reach out, grab, and lift it to the mouth. These syntagms, formed early in development, are gradually suppressed over the course of time via neural inhibition mechanisms, but there are clinical cases of neural damage where these suppression or inhibition mechanisms are damaged, and the automatic actions take place involuntarily in the presence of an object—we can't restrain ourselves from reaching for and lifting the spoon when we see it. While this level of grosser action on an object is only one form of the range of actions or transformations via action envisioned by Noë, the possibility of numerous suppressed syntagms, all semi-active or resonating so to speak as we view a scene with its objects, gives a glimpse of what the dynamic state of the brain in vision might involve. We should note finally that the concept of a linear "stream" of information flow (in the dorsal or in the ventral) is likely simplistic. Again, rather, we are talking about dynamic feedback (Gibson's resonance) among areas and ultimately very concrete dynamics.

THE BIOCHEMICAL BASIS OF THE SCALE OF TIME

The vision of the brain as involving a very concrete, bioelectronic dynamics was starting at the same time (at least) as Turing and the Information Processing model of the brain. In 1941, the biochemist, Szent-Gyorgyi, published a paper: "Towards a New Biochemistry?" [Szent1941]. At the beginning of the article, Szent-Gyorgyi noted that the study of crystals and metals has revealed the existence of a different state of matter. If a great number of atoms are arranged with regularity in close proximity, for example, in a crystal lattice, the terms of the single valency electrons may fuse into common bands. The electrons in this band cease to belong to one or two atoms only and belong to the whole system.

The electrons are normally stuck in the lowest orbit, and cannot do work, but supposing that one of these electrons is raised by the absorption of energy to a higher level, coming to excited state where it will move and transport its energy freely, it becomes impossible to say to which atom the excited electron belongs. Thus, he notes, the whole system can be looked upon as activated. By falling back to the lower level, the electron will give off its excess energy and perform work in a place more or less distant from that of the absorption of energy.

Szent-Gyorgi here gives a vision of something analogous to superconductivity, at minimum, a very highly conductive mass. In 1965, Snyder & Merril keyed off Szent-Gyorgi, noting some problems with a very widely held theory that LSD "is like [mimics] serotonin" and thus "interferes" in brain processes, causing hallucinations [Snyder1965]. They noted that an electronic, or "submolecular" hypothesis for the psychotropic actions of the drug has been proposed by Karreman, Isenberg, and Szent-Gy6rgyi, and that these researchers had performed molecular orbital calculations for chlorpromazine, LSD, and serotonin, concluding that these drugs were potent electron donors.

Molecular orbital calculations were made for several series of hallucinogenic drugs, and they examined the relationship between electronic configuration and hallucinogenic potency for a variety of phenylethylamine, amphetamine, and tryptamine derivatives, and for LSD. The prime factor they identified was "the energy of the Highest Occupied Molecular Orbital" (HOMO). This is a relative measure of the ability of an electron in the highest occupied molecular orbital of a compound to be transferred to an acceptor molecule. The greater the HOMO energy, the greater will be the propensity of a molecule to donate electrons. The highest of the ones they looked at was Mescaline (+0.53). LSD was not looked at, albeit it is known as even more potent. They noted Karreman et al., who calculated HOMO energy for the complete LSD molecule, obtaining a value of +0.218 A units—a HOMO far more energetic than any of the compounds examined in Snyder and Merril's study (the lower the value, the more potent as a donor).

Kang and Green, in 1970 [Kang1970], looked at this more deeply and complexified the HOMO thesis a bit. Basically, they saw that structural differences (stereochemical) in these substances also made a difference in effects, in potency. One could have a high HOMO, but a structural characteristic nullified this. So, another factor, likely structural, was seen to be involved.

Current Models of LSD's Action

Fast forward to a landmark study in 2016 by Carhart-Harris et.al [Carhart2016] representing the current state of the art in thinking on LSD and how it works in the brain. Complementary neuroimaging methods revealed that several brain regions were highly activated by LSD (at a dosage of 75 mcg) well beyond resting state levels. Increased blood flow to the visual cortex and a "greatly expanded" functional connectivity profile of the primary visual cortex was observed, together with increased connectivity among brain regions associated with vision, attention, movement, and hearing (i.e., the perceptual-motor regions). Decreased connectivity was observed between the parahippocampus and retrosplenial cortex, regions associated with maintenance of the sense of self.

Carhart-Harris et al. (2014, 2016) proposed an "entropic" framework for the psychedelic state: LSD breaks down or disrupts the organization of the brain, the normal intercommunication among areas, increasing with dose, descending to greater disorder (greater entropy). The prime focus of this proposal was the default mode network (DMN) with its subsystem, the medial temporal lobe (MTL). If disrupted, our tie to reality is degraded. The tie to the objective is maintained by the fine balance of certain major systems, for example, the DMN/MTL—and these are part of that breakdown noted under LSD [Carhart2014] [Carhart2016].

Shortly thereafter, another landmark study, Preller et al. [Preller2019] investigated the effects of 100 mcg LSD on the thalamus filter system. The structure of interest was the cortical-striatal-thalamo-cortical (CSTC) feedback loop, and the nodes of interest were the posterior cingulate cortex (PCC), the ventral striatum (VS), the thalamus, and the temporal cortex. The latter is associated with memory and "processing emotional and social information." LSD was proposed to create an abnormal flow of information into the cortex and to result in cognitive disruption, dreamlike experience, and ego dissolution.

Preller et al. noted the PCC is associated with arousal, awareness, and control of the balance between internally and externally directed thought. Failure to suppress activity of the PCC is therefore associated with the intrusion of internal mentation or an "overload." Importantly, the PCC is a core hub receiving information from other regions of the DMN, and for example, the medial prefrontal cortex, and angular gyrus.

Any tie here, in these studies, to Szent-Georgy, Snyder and Merril, and the HOMO concept as a factor in LSD's operation is gone. One can argue, of course, that this is because of the focus on both neuroimaging, the influence of the information processing metaphor, and the concept of "information flow," but it is likely more than that. Carhart-Harris et al. [Carhart2016], curiously, employ the term "perception" but once, this in discussing a psychedelic that brings about large changes in perception. This de-emphasis is likely sourced in the ubiquitous assumption of indirect realism and its concept that conscious experiences, including conscious perception and our experiential image of the coffee cup on the table, are simply *generated* by the brain. This is the "default mode" of current cognitive theory. In this framework, *there is no difference* between internal imagery or hallucinations and perception. This de-emphasis on perception is perhaps the cause of the subsequent de-emphasis of one of Carhart et al.'s main findings noted previously, namely, the perceptual-motor regions lighting up, with "increased communications." This finding does not necessarily fit with disorganization or an increasing entropy in/of brain regions, and the authors opted to assess the measure of cerebral blood flow to the visual cortex as unreliable, and essentially argued these findings are to be ignored.

Tying LSD to Perception

A very basic model at the origin of the energy models and the theory of form perception as a function of velocity flows was the Reichardt filter [Reichardt1959], an abstract neural filter with detectors and neural connections that will register the value of a velocity flow (Figure 3.10). An edge of a rotating cube could be moving by the filter, or in the case depicted, a fly's wing is moving by. This simple filter, however, is all we need to illustrate the subject of changing the biochemical substrate involved in specifying the scale of time.

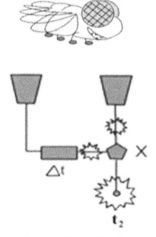

FIGURE 3.10. The Reichardt filter. The filter will register the velocity of the wing of the fly or of a heron passing by the two detectors (although the filter displayed is only "tuned" to detect motion in one direction, left-to-right) Multiple filters would be needed, tuned for the many possible directions of motion.

The two detectors are at fixed distance (d), and the Δt registered between t_1 (when the heron's wing first passes the leftmost detector, and t_2, when the wing subsequently passes the rightmost detector, allows computation of the velocity, $v = d/t$. This is good for a nice slow heron-wing. But assume a fly, where the fly's wings are moving extremely fast, so fast that multiple passes occur between the two detectors before the neural signal which converges at "X" traveling along the Δt neuron, gets to register. In other words, we're going to have a *blur* – the standard blur, say, of our normal scale of "buzzing" flies. It all depends on the transmission speed of those neurons, the transmission speed being the propagation rate of the *action potential*—the electrical change running along the neurons. This becomes the question: how can we speed up the action potential? Or, put differently, how can we speed up the neurons to turn our perception of the "buzzing" fly into that of a heron-like fly slowly flapping those wings?

What is the action potential (AP)? The neuron (Figure 3.11) is bathed, inside and outside, in a chemical, watery, liquid soup. The neural membrane separates two phases of electrolytes in the soup. A neurotransmitter (NT) hits the receptors at the start of the neuron (left end), and this stimulates a sodium channel: NA+ pours in. This starts the "action potential." A wavelike series of changes of the membrane potential is passing down the neuron, from channel to channel, from -70mV to +30mV, and back, with refractory periods (where nothing can happen). So, the voltage change (AP) passes "down" the neuron. That is, the electrolytes move from one side of the membrane to the other. The speed of this electrical flow is considered invariant (always the same speed of flow), and "all or none" (it happens, or it doesn't). This is the universally accepted view.

Near the terminal (at the right end) of the neuron, the AP opens a voltage gated CA2+ channel. The CA2+ floods in, binding to a "bubble" of NT, where such a bubble would hold the serotonin (which is LSD is thought to mimic). This releases the NT across the synaptic gap. The serotonin NT hits a receptor for serotonin at the start of the next neuron, where one such receptor is the 5-HT_{2A}, among others. This stimulates opening a sodium channel, starting the action potential down the neuron chain (from neuron A to neuron B, then down neuron C, etc.).

In the context of this AP-mechanics, LSD is called an "agonist," affecting several types of serotonin receptors, and for dopamine. *Agonist* is the term used in more recent studies. In standard terms, an agonist (a) acts like another NT, (b) increases the release of a NT, (c) opens a channel for a NT, and (d) blocks an enzyme that cleans up a NT (which means more of the NT is left sitting around).

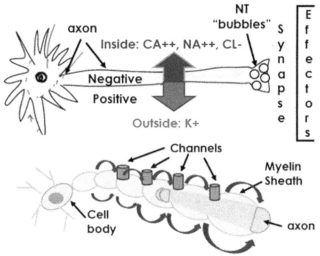

FIGURE 3.11. The neuron and the action potential. The cylinders represent the (conventional) gated channels bringing inside, for example, potassium, K+.

There are couple of puzzles in this general view of the AP. The first is simply, how does LSD work as an "agonist;" what are the mechanics? One can see statements such as, "LSD is a serotonin agonist that mimics the effect of serotonin at your brain cells but can produce a greater 'activation' than serotonin alone" [Bluelight], (or equivalently [WikiLSD]). But why, how, the "greater activation"? Secondly, and more crucially at this point of the discussion, why, in this view of the AP, is the system oscillating or in phase with itself (i.e., in some rhythm like the theta rhythm, the alpha rhythm)? Heat dissipation demands that a watery electrolyte proceeds to high entropy very quickly! No known electrical device reabsorbs heat; heat is only ever dissipated by electricity, not reabsorbed. You somehow need the "solidness" of a pendulum where you can give it a very low energy input and have it oscillated without being overwhelmed by heat fluctuation.

This brings us to the *Association-Induction* hypothesis (a different "AI") of Gilbert Ling [Ling1984]. We are going to replace the watery chemical soup. Visualize the brain as a big block of gelatin, in fact, a protein, one that keeps water in a highly, highly organized state. (Gelatin is mostly water, yet it behaves as a solid.) The water is "adsorbed" onto the gelatin; it is not in a free state. Adsorption occurs when a liquid, gas, or dissolved solid is adhered to the surface of the adsorbent. The water is "associated" with the brain, as well—again, not free, just as it is in gelatin. (In studies on meat, water cannot be centrifuged out even at very, very high speeds.) The electrolytes also—not free.

Ling demonstrated exhaustively that all positive ions (like positive potassium) are bound to their opposite charge on the cell. They are "associated" (the "association" part of his AI hypothesis), not just sitting there, not floating in space, not waiting to be "channeled" to either side of the membrane as the AP conventional framework views things.

- Conventional picture: Membrane with channels, bathed in electrolyte, in a watery soup
- Ling/AI: A gelatin onto which water and ions are tightly associated

This is why the brain can oscillate: The water and ions are in such a rigid state that they overwhelm heat dissipation and thermal fluctuation, like gears, a pendulum, or a watch. The "gears" are made of the protein/water/electrolyte mixture, which is in a highly organized, solid condition. Now we are approaching why there can be theta cycles and alpha cycles and why, for example, the firings of "place cells" in the hippocampus, as rats run through mazes (Cf. [Buzsáki2018]), are phase-locked in cycles.

The physical state of a gelatinous system can be modified by its surroundings. Try making Jell-O with vinegar added: Depending on how much vinegar you add, the Jell-O may not solidify. It may form many separate "orbs" surrounded by liquid water. This is key: the "adsorption" of water and ions is conditional on the state of the gelatin, and it is "cooperative." At a certain acidity, all the water will be free (unbound to other molecules), therefore the water "cooperates" going from one state to another based on the acidity of the Jell-O.

According to Ling, the acidity of this gelatinous mixture is under the control of a small number of molecules: the *cardinal adsorbent*. When ATP is present, the fixed negative charges of the living cell prefer potassium (K+). When the neuron "fires," the ATP is lost. Now the fixed negative charge does not prefer potassium or sodium. The sodium will come rushing in, or on the conventional picture, "channels are opening." Rather than seeing an individual neuron, the whole brain—with our hippocampus and that of Buszáki's rats also—can be viewed as a gelatinous mass. Under the control of ATP and other important molecules, it goes back and forth, from preferring to adsorb potassium and highly organized water, to having no preference for either. The generation of ATP by metabolism is therefore the small energy input, like pushing a pendulum.

When the electrolyte "leaves" the cell, the stuff outside the cell between the neurons is not meaningless. It (the space) is coordinating the whole mass. This is why Thoke and colleagues [Thoke2018a] showed that all billion yeast cells in a culture are in phase with each other, so long as the density is high enough. The space between the neurons is shared and coordinates the phase locking. It can be seen as one large, gelatinous mass.

Why, before we go on, in the conventional view, the "membrane channels" and "pumps?" The key here is that these work on the basis of ATP as well. So, if we take a single "pump" and study it in solution separates from the rest of the cell (a key methodology of biochemistry), we may come to the conclusion it is "pumping across a membrane and that it is powered by ATP." However, this is a leap of faith since the single pump itself will act like a gelatinous colloid; we are really just seeing a very small version of how the entire cell, indeed the entire brain, would act. In the presence of ATP the "pump" prefers one ion. Otherwise, it has no preference.

Increasing the Velocities Underlying Perception

How to speed up these phase locked cycles? "Speed" is the concept of chemical velocity, expressed in one form by the Arrhenius equation [Laidler1996]:

$$k = Ae^{(-E_a/RT)} \tag{1}$$

Here A is a constant, R is the universal gas constant, T is the absolute temperature, and E_a is the activation energy or rather, the minimum energy required to initiate a chemical reaction. Given the T in the equation, an increase in temperature increases chemical velocity, and Hoagland [Hoaglund1966], who noted this equation, also noted the slowing of experienced time is common to fevers. This temperature dependence of neural conduction velocities is common to many species [Yu2012] [Fillafer2013].

In the biochemical framework described by Robbins and Logan [RobLog2022], two of the major features are:

- LSD and its receptors influence the frequency of biochemical oscillations near-equilibrium via an electronic effect, for example the relaxation of cell water in phase with chemical products of metabolism [Thoke2018b] [Begarani2019].

- Gilbert Ling's association-induction hypothesis and the Yang-Ling isotherm [Ling1984] describe the influence of an effector via electronic induction, upon the cooperative interaction of many other ions and molecules.

Many observations appear consistent with the hypothesized increase in rate processes, including: the temperature changes (noted previously) that are straightforwardly related to conduction velocity, the repeated observation of an alpha wave frequency increase by LSD [Fink1969] [Carhart2016], the increased BOLD signals and CBF in visual regions of the human brain on LSD [Roseman2016], a comprehensive increase in the frequency of model nervous systems by LSD [Wright1962], and more.

As indicated earlier, the association-induction hypothesis is a statistical mechanical model for the cooperative adsorption of ions, water, and other molecular species under the control of a smaller number of "cardinal" adsorbents [Ling1984]. The Yang-Ling adsorption isotherm is informed by next-neighbor interaction energies and is general [Ling1984] [Thoke2018a], recapitulating noncooperative as well as cooperative effects, such as the binding of oxygen to hemoglobin that is usually described by the Hill equation. The isotherm is [Ling1984]:

$$[p_i]_{ad} = \frac{[f]}{2}\left[1 + \frac{e_0 - 1}{\sqrt{(e_0 - 1)^2 + 4e_0 e^{(y/RT)}}}\right] \tag{2}$$

Here "$[p_i]_{ad}$" is the concentration of the ith species adsorbed onto [f] number of sites, as opposed to an alternate species j. R and T are the universal gas constant and absolute temperature, respectively, and y is the important next-neighbor interaction energy or rather, the free energy required to create one mol of ij neighboring pairs such that a change from *iii* → *iji* entails an energy of *-y* [Ling1984]. Note that the ultimate term of the denominator is of the same form as the activation energy in Equation 1, and 5-hydroxytryptamine (5-HT) receptor binding (a key receptor involved with LSD) is known to require the cost of such a barrier [Noller1982]. "e_0" is the concentration ratio of nonadsorbed i and j in solution, multiplied by their intrinsic equilibrium constant K [Ling1984].

In [RobLog2022] it was argued that the exchange of one cardinal adsorbent for another, such as ATP for ADP [Thoke2018a], or in this case serotonin for the other 5-HT receptor analogs, may shift the next-neighbor interaction energy in Equation 2 from y to y_c, where y_c is the next-neighbor interaction energy in the presence of the new cardinal adsorbent. In some cases, the replacement of one cardinal adsorbent by another will lower y in Equation 2 to such an extent that the system becomes auto-cooperative, that is, each adsorbed species of a particular kind increases the probability of further adsorption for that species.

This framework is dependent upon a relatively long-range electronic effect induced by LSD and propagated via its receptor proteins [Ling1984] [Begarani2019]. The "induction" aspect of the association-induction hypothesis is the equivalent physical mechanism by which parameter y in Equation 2 is lowered.[3] It was noted in the introduction to this topic, that early reports, keying off Szent-Gyorgi, indicated a strong correlation between the subjective effects of psychedelic compounds and their electron orbital energies [Snyder1965] [Kang1970] [Domelsmith1978a] [Domelsmith1978b]. The import of these data has been neglected; LSD, DMT, and Psilocybin are simply lumped together, therefore, the *independence* of structural and chemical features across these

---

[3.] Importantly, the fluorescence spectrum of LSD shifts when bound to its receptor—but not to controls—implying that delocalizations of electrons are coincident effects [Shih1977].

drugs is that considered relevant to current proposals of biochemical mechanism [Perez-Aguilar2014].

The "inductive index" of Chiang and Tai [Chiang1963] has been adapted for the purpose of quantifying the propensity of R groups to donate or withdraw electron density [Lara-Popoca2020], and in [RobLog2022] a modified inductive index was determined via extrapolation for all 20 of the standard amino acids (cf. [RobLog2022], Table S1). The important receptor 5-HT_{2A} was found to have the highest mean inductive index and to be the most enriched with electron withdrawing R groups (Figure 3.12) among amino acids in the primary sequence of 16 high affinity LSD receptors. The 5-HT_{2A}R is the common target of all psychedelic compounds [Perez-Aguilar2014], and the significant, electron withdrawing character of its amino acids appears to be a very significant property.

Protein	Amino Acids LSD receptors	Mean Inductive Index	Proportion EWG
5-HT_{1B}R	390	-47.7	.431
5-HT_7R	479	-53.52	.413
...			
D3	400	-43.13	.448
Alpha-2AR	465	-62.98	.385
...			
5-HT_{2B}R	481	-31.79	.47*
5-HT_{2A}R	471	-27.2	.48**

FIGURE 3.12. Snippet of Table 1 from [RobLog2022]. "EWG" refers to amino acids with electron withdrawing R groups (Table S1 in Robbins and Logan). An asterisk represents $p < .05$ and two asterisks represent $p < .01$.

This, hopefully, is a sufficiently descriptive view of the biochemical framework underlying the action of LSD described in [RobLog2022]. In summary, LSD acts as a cardinal adsorbent, like ATP, but more powerful. The oscillation of the entire mass (or at least the portions LSD affects) is sped up. The frequency of the cycles—theta, alpha—increases. The *specified scale of time changes*:

- The buzzing fly becomes heron-like.
- The spoon circles more slowly in the coffee, the liquid swirls slowly down, the mouse is moving by more slowly.
- Speech—as heard—slows.
- Colors, because the brain is a bit closer now in vibratory rhythm to the oscillating fields that support colors, are more "vibrant."

▪ Visual acuity increases. This too is a very common report of LSD effects, and in Robbins and Logan, arguments are given as to why this might be true when viewed in terms of the transformation of spatial frequency to temporal frequency and vice versa.

This is obviously a concrete dynamic, not abstract symbol manipulation. The contrast between the biochemical framework described and the prevailing neurocomputational view of the action potential should be noted here. The action potential, as we saw, is generally viewed as a one-way chemical flow down a neuron, proceeding from neurotransmitters at the initiating dendrites, and via voltage-gated channels along the axon, to the release of neurotransmitters at the axon terminal, the strength of which is taken as a connection strength to adjacent neurons. Given multiple neurons, the matrix of connections serves as a conceptual basis for neural net computations. In the oscillatory framework described here, the action potential is one-half of one oscillation, and a particular neuron is one oscillator within a medium containing billions of oscillators, of which the rate of oscillation is affected by electronic induction.

One can see this conception is edging closer to supporting the view that the brain itself is a complicated, continuously modulated, very concrete waveform. There are some other implications: The neuron is normally conceived as a one-way flow; at the end of the flow, it connects to another with a "connection strength." Another stimulus occurs: the neuron fires (one-way) again. This framework (neglecting the need for back-propagation and the derivative computations now also involved) is, as noted, the basis of the neural net conception. But the neuron is phase-locked in a cycle (e.g., the theta) —it (with its AP) is an oscillator. The cell spikes are relative to the phase, something like interlocking gears. This oscillator view, we would submit, is not the neural net concept on which neural-net AI is based.

LSD-MODULATED PERCEPTION AS OBJECTIVE

This discussion returns us to that very standard framework of indirect realism in which current cognitive science and AI works. If there were a short phrase that describes the common theoretical stance toward the effect of psychedelics on brain function, it would be: "*All* is hallucination." The various perceptual effects that are said to characterize, for example, LSD, things like "time distortion," apparently enhanced visual acuity, pulsating objects, vibrant colors, are taken to be in the same category as hallucinations in general, to include bugs crawling out of the wall, Hinton's "pink elephants" (see Chapter 1) or other strange visual displays. In fact, not too much attention is paid in either the empirical studies or in the theoretical work as to whether the subject has their eyes open or closed—both states are considered nearly equivalent. This approach is anchored in the concept that the brain *generates* the image of the external world; in fact, it generates *all* images—internal images or imaginative images as well as the image of the coffee cup out there on the kitchen table. It is, in other words, an implicit assumption of indirect realism: the world out there—the coffee cup with the stirring spoon—is indeed objective, but the image thereof is generated by the brain and is somehow within the mind or "mental space" along with all other "internal" images. But the difficulty is that neuroscience has no model of how the brain and its neural mass creates or produces images—it is simply a faith that it does.

This "all is hallucination" indirect realist framework has been the stance of major LSD researchers such as Carhart-Harris [Carhart2014], [Carhart2016], and [Preller2019]. Although characterization of structural features important for b-arrestin signaling (Wacker et al., [Wacker2013] [Wacker2017]) was a crucial advance (only discussed indirectly here re the long-lasting effect [for hours] of LSD), a unifying theme for LSD biochemistry is absent. This is partly because the effects suffer from being vaguely related, for example, effects of LSD upon scale independent measurements, the correlation between electron orbital energies and subjective effects, and so on, but it is, as well, heavily because of the indirect realist position. Indirect realism and its hallucinations can be made consistent, *after the fact*, with *any change* to biochemical rate processes (faster, slower, more erratic) and can be made consistent with biased and unbiased signaling, equally. In contrast, the Bergson-Gibson perceptual model is only consistent with increased rate processes of the relevant systems and, given the applicability of the association-induction hypothesis, with a relatively powerful electronic effect. But in the grips of indirect realism, all future biochemical data—no matter what their intricacy—will remain "related" to hallucinations.

So, the stance of this model is quite different, but let us stress again, we are only focused on LSD here, not other psychedelics such as psilocybin or DMT, the latter seeming to have a few very commonly experienced, one almost wants to say "objective" experiences/dimensions, one of these being a dimensional-realm, often reported, where one is examined by aliens [Strassman2001], and so on. These drugs employ different mechanisms and have their own particular phenomenal effects. The stance here is that at least within a certain LSD-dosage limit (after which the "internal" effects become too overwhelming, e.g., the "flood" of memories) the perceptual effects of LSD are veridical. To rephrase:

- These altered perceptions are veridical, at least as veridical as perception can be, for perception is always an optimal specification of the external world.
- These effects—fans slowing, flies slowing, conversations or speech slowing, colors more vibrant—are timescale transformations of the specification of the external world, and a function of the chemical dynamics involved.
- The visual acuity changes are a natural, correlated effect (for spatial frequency—as in a grid of black/white bars or the big E of the vision-test card—can be related to temporal frequency).
- These are not hallucinatory effects, nor a function of the brain's generation of hallucinatory images.

In other words, these are the natural effects of modulating the biochemical base involved in specifying the scale of time.

Theorists like Carhart-Harris and Preller have been focused on LSD's effect on neural mechanisms, for example, a thalamic gateway, such that a "flood of information" is now released into the brain, overwhelming our normal tie to reality (i.e., their assumed indirect realist form thereof), and hence the even greater than normal level of hallucination (remembering that in the indirect view even normal perception is a form of hallucination). A correlated take on this is, as dosage increases, the normally tight tie of perception to action in the environment (thus the re-entrant interaction of the visual area with the motor areas) is increasingly loosened, disturbed. In this we approach Bergson's suggestion that the brain be seen as a form of "valve," where certain experiences from the past enter into the present state of consciousness according to the current configuration of the action systems, or as we've already explored, according to the resonating pattern reflective of the invariance structure of external events and the intrinsic tie to action systems involved. Only those events that "fit" the resonance/action-system configuration "get through" into the present state.

But Bergson's "valve-like" resonating, reconstructive wave-brain, specifying the coffee cup "out there," is now also serving as the opening/closing mechanism to the entire 4-D extent of our experience, for in the indivisible transformation of the holographic field, we are now viewing the human as a 4-D being, the body being the 3-D "leading edge" thereof. This is part of the problem of memory that we will explore further in subsequent chapters.

AI: THE PROBLEM OF ECOLOGICAL INTERACTION WITH THE ENVIRONMENT

Let's sum up the hurdles as seen thus far (for there are more to come) that an AI faces if it is to interact—truly, as humans do—with the ecological world:

- The human brain is being driven, as a resonating field, by the invariance structures defining external events. In its resonance, it symmetrically reflects the on-going event structure.

- This resonance is simultaneously a modulated reconstructive wave embedded within a holographic field and specific, now as an *image*, to sources (events) within the field.

- As this brain is intrinsically embedded within, thus intrinsically participating in, the invisible transformation of the holographic field, it is enabled to be specific to the past of the field, and it is not relying on any form of short-term memory to store "instants" or "states" of the past event such that it can be specified.

- The specified event is simultaneously the reflection of the possible (virtual) action of the organism on the world. As such, the event is inherently meaningful; it is already a "symbol," and the symbol is inherently grounded (although we shall examine the greater complexity behind symbol grounding in Chapter 4, "Retrieving Experience: Implicitly and Explicitly").

- As the holographic field, due to its indivisible and melodic transformation, is inherently qualitative, the specification is already qualitative—it can be to "mellow" violins, ever so gently waving curtains, or reddish sunsets.

- It is a multimodal specification, that is, specific to the same event within multiple modalities within the external, holographic field, where the (amodal) invariants are coordinate, simultaneously, across these modalities.

- The specification determines the scale of time at which the external field is perceived. Again, the specification as qualia becomes intrinsic—buzzing flies or heron-like flies. These dynamic images are simultaneously reflective of the possible (virtual) action at this scale.

- The dynamics of this time-scale specification can be modified at the biochemical level.

- This all involves very concrete, biochemical dynamics. It is in no way achieved by software, by symbol manipulation, or by *software* engineering.

The implications of the "multimodal specification" point should be registered. It is thought that the way to AGI is to give a robot a camera to register the visual world, a microphone (for the acoustics), pressure sensors (for the kinesthetics) and robotic hand arms (for action) with perhaps an analog neural net (e.g., [Hinton2023], also [Chalmers2022]). But all this input (from the various modalities) is being *transduced to* a single, *homogeneous* medium—the computer memory, and ultimately to symbolic representation therein (whether by weights or by symbols), therefore, to the abstract space of the classic metaphysic. The different modalities are now lost, and not recoverable—the representation is homogeneous. This is not what is going on in the Bergson-Gibson "device"—the specification being indeed to each different modality of the event within the external field, not "within" the device.

This is a beginning list of hurdles. The "storage" of experience is now entirely different, and as well, cognition.

REFERENCES

[Begarani2019] Begarani, F., D'Autilia, F., Signore, G., Del Grosso, A., Cecchini, M., Gratton, E., Beltram, F., & Cardarelli, F. "Capturing metabolism-dependent solvent dynamics in the lumen of a trafficking lysosome," *ACS Nano* (2019): *13*(2), 1670–1682. https://doi.org/10.1021/acsnano.8b07682

[Bergson1889] Bergson, H. *Time and Free Will: An Essay on the Immediate Data of Consciousness.* London: George Allen and Unwin Ltd., 1889.

[Bergson1896] Bergson, H. *Matter and Memory.* New York: Macmillan, 1896/1912.

[Bergson1907] Bergson, H. *Creative Evolution.* New York: Holt, 1907/1911.

[Bluelight] LSD and Serotonin | Bluelight.org

[Bohm1980] Bohm, D. *Wholeness and the Implicate Order.* London: Routledge and Kegan-Paul, 1980.

[Boring1942] Boring, E. G. *Sensation and Perception in the History of Experimental Psychology.* New York: Appleton, 1942.

[Brown2015] Brown, H., & Friston, K. "Free energy and illusions: The Cornsweet Effect," *Frontiers in Psychology* (2015): *3*(43), 1–13.

[Buzsáki2017] Buzsáki, G. *The Brain from Inside Out.* Oxford: Oxford University Press, 2017.

[Buzsáki2018] Buzsáki, G., & Tingley, D. "Space and time: The hippocampus as a sequence generator," *Trends in Cognitive Sciences* (2018): *22*(10), 853–869. https://doi.org/10.1016/j.tics.2018.07.006

[Carhart2014] Carhart-Harris R. L., Leech, R., Hellyer, P., Shanahan, M., Feilding, A., Tagliazucchi, E., Chialvo, D., & Nutt, D. "The entropic brain: A theory conscious states informed by neuroimaging research with psychedelic drugs," *Frontiers in Neuroscience* (2014): 8(Article 20), 1–22. https://doi.org/10.3389/fnhum.2014.00020

[Carhart2016] Carhart-Harris, R. L., Muthukumaraswamy, S., Roseman, L., Kaelen, M., Droog, W., Murphy, K., Tagliazucchi, E., Schenberg, E., Nest, T., Orban, C., Leech, R., Williams, L. T., Williams, T. M., Bolstridge, M., Sessa, B., McGonigle, J., Sereno, M., Nichols, D., Hellyer, P., Hobden, P., Evans, J., Singh, K., Wise, R. G., Curran,

H. V., Feilding, A., & Nutt, D., "Neural correlates of LSD experience revealed by multimodal neural imaging," *PNAS* (2016): *113*(17), 4853–4858.

[Chalmers96] Chalmers, D. J. *The Conscious Mind*. New York: Oxford University Press, 1996.

[Chalmers2022] Chalmers, D., "Are Large Language Models Sentient?," October 2022: https://www.youtube.com/watch?v=-BcuCmf00_Y

[Chiang1963] Chiang, M.C., & Tai, T.C.. "A quantitative relationship between molecular structure and chemical reactivity," *Sci. Sin* (1963): *12*, 785-867. https://sioc-journal.cn/Jwk_hxxb/EN/abstract/abstract342922.shtml

[Churchland1994] Churchland, P. S., Ramachandran, V. S., & Sejnowski, T. J. "A critique of pure vision," In C. Koch & J. Davis (Eds.), *Large-scale Neuronal Theories of the Brain. Cambridge*: MIT Press, 1994.

[Clark2008] Clark, A. *Supersizing the Mind: Embodiment, Action and Cognitive Extension*. Cambridge: Oxford University Press, 2008.

[Domelsmith1978a] Domelsmith, L. N., & Houk, K. N. "Photoelectron spectroscopic studies of hallucinogens: The use of ionization potentials in QSAR," in G. Barnett, M. Trsic, & R. E. Willette (Eds.), *Quantitative Structure Activity Relationships of Analgesics, Narcotic Antagonists, and Hallucinogens* (pp. 423-440). Rockville, MD: National Institute of Drug Abuse, 1978.

[Domelsmith1978b] Domelsmith, L. N., Munchausen, L. L., & Houk, K. N. "Lysergic acid diethylamide: Photoelectron ionization potentials as indices of behavioral activity," *Journal of Medicinal Chemistry* (1978): *20*(10), 1346–1348. https://doi.org/10.1021/jm00220a024

[Fillafer2013] Fillafer, C., & Schneider, M.F. "Temperature and excitable cells testable predictions from a thermodynamic perspective," *Communicative and Integrative Biology* (2013): *6*(6). https://doi.org/10.4161/cib.26730

[Fink1969] Fink, M., "EEG and human psychopharmacology," *Annual Review of Pharmacology* (1969): *9*, 241–258. https://doi.org/10.1146/annurev.pa.09.040169.001325

[Fischer1966] Fischer, R. "Biological time," in J. T. Fraser (Ed.), *The Voices of Time*. New York: Braziller, 1966.

[Gallese2007] Galese, V. "The "Conscious" dorsal stream: Embodied simulation and its role in Space and Action conscious awareness." *Psyche* (2007): *13*(1), 1–20.

[Gibson1966] Gibson, J. J. *The Senses Considered as Visual Systems.* Boston: Houghton-Mifflin, 1966.

[Gibson1975] Gibson, J. J. "Events are perceivable but time is not," in Fraser, J. T., Lawrence, N. (Eds.), *The Study of Time II.* Springer, Berlin, Heidelberg, 1975.

[Hoaglund1966] Hoagland, H. "Some biochemical considerations of time," in J. T. Fraser (Ed.), *The Voices of Times* (pp. 312–329). New York: Braziller, 1966.

[James1890] James, W. *Principles of Psychology.* New York: Holt and Co., 1890.

[Kang1970] Kang, S., & Green, J. P. "Steric and electronic relationships among some hallucinogenic compounds," *PNAS* (1970): *67*, 62–67.

[Laidler1996] Laidler, K. J. "A glossary of terms used in chemical kinetics, including reaction dynamics, *Pure and Applied Chemistry.* (1996): *68*(1), 151. https://doi.org/10.1351/pac199668010149

[Lara-Popoca2020] Lara-Popoca J., Thoke H.S., Stock R.P., Rudino-Pinera E., & Bagatolli, L.A. "Inductive effects in amino acids and peptides: Ionization constants and tryptophan fluorescence," *Biochem. Biophys. Rep.* (2020). https://doi.org/10.1016/j.bbrep.2020.100802

[Lehar2001] Lehar, S. "Gestalt Isomorphism and the Primacy of the Subjective Conscious Experience: A Gestalt Bubble Model" (2001): Available online at: http://www.iro.umontreal.ca/~ostrom/dift6095/Gestalt_reader/Gestalt%20Isomorphism.htm

[Ling1984] Ling, G. *In Search of the Physical Basis for Life.* New York: Plenum Press, 1984.

[Lynds2003] Lynds, P. "Time and classical and quantum mechanics: Indeterminacy versus discontinuity," *Foundations of Physics Letters* (2003); *16*(4), 343–355.

[Merleau1962] Merleau-Ponty, M. *Phenomenology of Perception.* (C. Smith, Trans.) London, Routledge, 1962.

[Milner1998] Milner, A., & Goodale, M. "The visual brain in action," *Psyche (1998):* 4(12), http://psyche.cs.monash.edu.au/v4/psyche-4-12-milner.html

[Nakamura1980] Nakamura, R. K. & Mishkin, M. "Blindness in monkeys following non-visual cortical lesions," *Brain Research* (1980): *188*(2), 572–577.

[Nakamura1982] Nakamura, R. K. & Mishkin, M. "Chronic blindness following non-visual lesions in monkeys: Partial lesions and disconnection effects," *Society of Neuroscience Abstracts* (1982): *8*, 812.

[Noë2004] Noë, A. *Action in Perception.* Cambridge, MA: MIT Press, 2004.

[Noller1982] Noller, H., Kienzl, E., & Riederer, P., "Coordination chemical aspects of receptor biochemistry," iIn M. Goldstein, K. Jellinger, & P. Riederer (Eds.), *Basic Aspects of Receptor Biochemistry* (pp. 45-54). Vienna, AT: Springer, 1982.

[Nourbakhsh2013] Nourbakhsh, R. *Robotic Futures.* Cambridge, MA: MIT Press, 2013.

[Perez-Aguilar2014] Perez-Aguilar, J. M., Shan, J., LeVine, M. V., Khelashvili, G., & Weinstein, H. "A functional selectivity mechanism at the serotonin-2A GPCR involves ligand-dependent conformations of intracellular loop 2," *Journal of the American Chemical Society.* (2014): *136* (45), 16044–16054.

[Pittenger1975] Pittenger, J. B., & Shaw, R. E. "Aging faces as viscal elastic events: Implications for a theory of non-rigid shape perception," *Journal of Experimental Psychology: Human Perception and Performance* (1975): *1*: 374–382.

[Preller2019] Preller, K., Razi, A., Zeldman, P., Stampfli, P., Friston, K., Vollenweider, F., "Effective connectivity changes in LSD-induced altered states of consciousness in humans," *PNAS* (2019): *116*(7), 2743–2748.

[Pribram1971] Pribram, K. *Languages of the Brain.* Englewood Cliffs, NJ: Prentice-Hall, 1971.

[Purves2010] Purves, D., & Lotto, B. *Why We See What We Do Redux: A Wholly Empirical Theory of Vision.* Oxford University Press, 2004.

[Quanta2023] "Physicists who explored tiny glimpses of time win the Nobel Prize," https://www.quantamagazine.org/physicists-who-explored-tiny-glimpses-of-time-win-nobel-prize-20231003/, 2023.[Raja2018] Raja, V. "A theory of resonance: Towards an ecological cognitive architecture," *Minds and Machines* (2018): 28(13), 29–51.

[Reichardt1959] Reichardt, W., "Autocorrelation and the central nervous system," In W.A. Rosenblith (Ed.), *Sensory Communication* (pp. 303-318). Cambridge, MA: MIT Press, 1959.

[Robbins2000] Robbins, S. E. "Bergson, perception and Gibson," *Journal of Consciousness Studies* (2000): 7(5), 23—45.

[Robbins2004] Robbins, S. E. "On time, memory and dynamic form," *Consciousness and Cognition* (2004): 13(4), 762–788.

[Robbins2006] Robbins, S. E. "Bergson and the holographic theory," *Phenomenology and the Cognitive Sciences* (2006): 5(3–4), 365–394.

[Robbins2013] Robbins, S. E. "Time, form and qualia: The hard problem reformed," *Mind and Matter* (2013): 11, 153–181.

[Robbins2014] Robbins, S. E. *Collapsing the Singularity: Bergson, Gibson and the Mythologies of Artificial Intelligence.* Atlanta: CreateSpace, 2014.

[Robbins2021] Robbins, S. E. "Bergson's Holographic Theory—60—Quantum Jumps—Not," October 16, 2023: https://www.youtube.com/watch?v=xvWmvCmiHDc&t=1850s

[Robbins2023] Robbins, S. E. "Gibson and time: The temporal framework of direct perception," *Ecological Psychology* (2023): 35(1), 31–50.

[RobLog2022] Robbins, S. E. & Logan, D. "LSD and perception: The Bergson-Gibson model for direct perception and its biochemical framework," *Psychology of Consciousness* (2022): 9(4), 305–335. https://doi.org/10.1037/cns0000302

[Roseman2016] Roseman, L., Sereno, M. I., Leech, R., Kaelen, M., Orban, C., McGonigle, Feilding, A., Nutt, D. J., & Carhart-Harris, R. L. "LSD alters eyes-closed functional connectivity within the early visual cortex in a retinotopic fashion," *Human Brain Mapping* (2016): 37(8). https://doi.org/10.1002/hbm.23224

[Searle2015] Searle, J. *Seeing Things as They Are: A Theory of Perception.* Oxford: Oxford University Press, 2015.

[Shih1977] Shih, J.C., & Rho, J., "The specific interaction between LSD and serotonin-binding protein," *Research Communications in Chemical Pathology and Pharmacology*. (1977): *16*(4), 637–647. https://doi.org/10.1016/0006-8993(78)90497-3

[Shröd1952a] Schrödinger, E. "Are there quantum jumps? Part I," *The British Journal for the Philosophy of Science* (1952): *3*(10), 109–123.

[Shröd1952b] Schrödinger, E. "Are there quantum jumps? Part II," *The British Journal for the Philosophy of Science* (1952): *3*(11), 233–242.

[Snyder1965] Snyder, S. H. & Merril, C. R. "A relationship between the hallucinogenic activity and their electronic configuration," *PNAS* (1965): *54*, 258–266.

[Strassman2001] Strassman, R. *DMT: The Spirit Molecule*. Rochester, Vermont: Park Street Press, 2001.

[Szent1941] Szent-Gyorgyi, A., "Towards a new biochemistry?" *Science* (1941): *93*(2426), 609–611.

[Taylor2002] Taylor, J. G. "From matter to mind," *Journal of Consciousness Studies* (2002): *9*(4), 3–22.

[Thoke2018a] Thoke, H. S., Bagatolli, L. A., & Olsen, L. F. "Effect of macromolecular crowding on the kinetics of glycolytic enzymes and the behavior of glycolysis in yeast," *Integrative Biology* (2018}: 10(*10*), 587–597. https://doi.org/10.1039/c8ib00099a

[Thoke2018b] Thoke, H. S., Olsen, L. F., Duelund, L., Stock, R. P., Heimberg, T., & Bagatolli, L. A. "Is a constant low-entropy process at the root of glycolytic oscillations?" *Journal of Biological Physics*, (2018): *44*(3), 419-431. https://doi.org/10.1007/s10867-018-9499-2

[Turvey1977a] Turvey, M. "Contrasting orientations to the theory of visual information processing," *Psychological Review* (1977): *84*(1), 67–88.

[Turvey1977b] Turvey, M. "Preliminaries to a theory of action with references to vision," in R. E. Shaw & J. Bransford (Eds.), *Perceiving, Acting and Knowing* (pp. 311–321), Hillsdale, NJ: Erlbaum, 1977.

[Turvey2019] Turvey, M. *Lectures on Perception*. New York: Routledge, 2019.

[Viviani1990] Viviani, P., & Mounoud, P. "Perceptuo-motor compatibility in pursuit tracking of two-dimensional movements," *Journal of Motor Behavior* (1990); *22*(3), 407–443.

[Viviani1992] Viviani, P. & Stucchi, N. "Biological movements look uniform: Evidence of motor-perceptual interactions," *Journal of Experimental Psychology: Human Perception and Performance* (1992): *18*(3), 603–623.

[Wacker2013] Wacker, D., Wang, C., Katritch, V., Won Han, G., Huang, X. P., Vardy, E., McCorvy, J. D., Jiang, Y., Chu, M., Yiu Siu, F. Y., Liu, W., Xu, H. E., Cherezov, V., Roth, B. L., & Stevens, R. C. "Structural features for functional selectivity at serotonin receptors," *Science* (2013): *340*(6132), 615–619. https://doi.org/10.1126/science.1232808

[Wacker2017] Wacker, D., Wang, S., McCorvy, J. D., Betz, R. M., Venkatakrishnan, A. J., Levit, A., Lansu, K., Schools, Z. L., Che, T., Nichols, D. E., Shoichet, B. K., Dror, R. O., & Roth, B. L. "Crystal structure of an LSD-bound human serotonin receptor," *Cell* (2017): *168*(3), 377–389. https://doi.org/10.1016/j.cell.2016.12.033

[Warren1984] Warren, W. H. "Perceiving affordances: Visual guidance of stair climbing," *Journal of Experimental Psychology: Human Perception and Performance* (1984): *10*(5), 683–703.

[Weiskrantz1997] Weiskrantz, L. *Consciousness Lost and Found*. New York: Oxford, 1997.

[Whitehead1929] Whitehead, A. N. *Process and Reality*. New York: Free Press, 1929/1978.

[WikiLSD] LSD - Wikipedia

[Wright1962] Wright, A. M., Moorhead, M., & Welsh, J. H. "Actions of derivatives of lysergic acid on the heart of Venus mercenaria," *British Journal of Pharmacology and Chemotherapy* (1962): *18*(2), 440–450. https://doi.org/10.1111/j.1476-5381.1962.tb01422.x

[Yu2012] Yu, Y., Hill, A. P., & McCormick, D. A. "Warm body temperature facilities energy efficient cortical action potentials," *PLoS Computational Biology*. (2012): *8*(4). https://doi.org/10.1371/journal.pcbi.1002456

CHAPTER 4

RETRIEVING EXPERIENCE: IMPLICITLY AND EXPLICITLY

LASHLEY'S RATS: ENGRAMS—MISSING

The coffee cup with stirring spoon, as an image, is being specified right where common sense says it is, within the external, holographic field. While the action systems, neural systems, and perceptual processing systems are intrinsically involved in this specification, our experience (the coffee stirring) is not occurring solely within the brain, therefore it cannot be solely "stored" there.

Bergson, we noted earlier, distinguished two forms of memory: (1) motor memory, and (2) experience. Motor memory, or "procedural" or "habit memory," he saw as a modification of the brain's neural networks, perhaps like a connectionist net in our current understanding. It is a mechanism that can be unrolled at will, for example, our tennis serve or the end result of our practice sessions of Chopin's Waltz in C# minor, where we can unroll this waltz on the piano at will. Experience (or episodic memory) is not stored in the brain, but rather, given the indivisible flow of time, of being, of experience, our experience is the 4-D extent of being. Every experience of practicing the waltz: the day we were grumpy, the day it was storming outside, the day the piano teacher was furious—all this is "in" the 4-D time-extent of our being. Our "past," as Bergson stated, is *what no longer can be acted upon*.

It will be asked immediately, "But what of the *loss* of memories, for example, in amnesia or aphasias, for example, loss from lesions or damage to the brain? Is this not the absolute proof that not only motor memory but also our experience is stored in the brain?" These losses indeed have been taken as *the* proof of brain-storage of experience, but the answer, as we will see, is that "loss" can equally be interpreted, not as damage or destruction to memories stored in the brain, but as damage to the brain's neural structures that support the complex wave-modulation patterns defining a specific reconstructive wave—to use the holographic framework—required to access and retrieve events in our 4-D being—our experience. Damage to the ability to support these modulatory patterns *looks* like the loss of *stored* (in the brain) experience. This subject will be developed shortly, but initially, we continue with motor memory.

In Bergson's *Matter and Memory* there is the indication that even in motor memory, more is involved than simple storage in the brain or neural net like modifications. This, we will see, was realized explicitly within Bergson's later discussion of "Dynamic Schemes" [Bergson1902], but this brings us not many years later (1929–1930) to Karl Lashley and his search for the "engram," that is, the search for the precise locale in the brain where a given motor memory is stored in the brain [Lashley1950], perhaps the sequence of actions required (if one is a rat) to make one's way through a maze.

Lashley was of course aware of Pavlov and his experiments with conditioning. Dogs, conditioned to hear a bell which was followed by feeding, eventually began to salivate on hearing the bell. For Pavlov, this was a "conditioned reflex." To many scientists of the time, this was termed "reflex arcs," like connecting wires in the then-current technology, namely, a telephone exchange. But Pavlov was reluctant to say this. He had discovered that the conditioning could survive massive surgical damage to the brain! Lashley too suspected the "reflex arc" analogy. He trained rats to respond in specific ways to a light, then nearly all their motor cortex was cut out. They could still perform as well as non-operated-on rats.Monkeys were trained to open boxes with latches, then were subjected to an operation wherein nearly all the motor cortex was tossed. After two–three months (when they could move around again), they opened the boxes promptly, without hesitation. So much for the motor cortex as the storage area for this memory. Now for the associative areas of the cortex: deep incisions were made destroying cross-connections, but still there was no problem with the task. When the cerebellum was removed, there was no problem with the task.

This led to Lashley's view that memories were *distributed* over the brain, or at least over a functional area. There is of course an alternative, namely, that such memories are not stored in the brain at all. Lashley came very near this hypothesis; he had already concluded the worst: *there are no learned, stored, engraved circuits.* Consider briefly what this implies for the then-in-the-future neural net architecture (or connectionist net or deep learning net). These nets are precisely the *fixed circuits* that Lashley denied. Take a net designed/trained to discriminate or categorize the digits 0–9. Perhaps the net has an input layer of 784 units (for a 28 × 28-pixel grid), two hidden layers of 16 units, and an output layer of 10 units (each representing a specific number, 0, 1, 2 … 9), thus, each output unit containing the probability of the number presented to the input layer. The output unit (of the ten) with the highest probability is taken as the number presented, for example, if the output unit for "8" holds the highest probability, the input number (i.e., pixel pattern) is classed as an "8." This, again, is a *fixed* architecture. After training, the connection weights learned, and the net is classifying the numbers correctly, suppose Lashley appears, ruthlessly severing boatloads of connections from the input units to the hidden layers, and a chunk of connections from the last hidden layer to the output layer. The learned weight classification scheme is destroyed, the net's classification ability is gone; it can no longer respond correctly. In other words, it is hard to avoid the conclusion that Lashley had *already performed* a rather ruthless experiment—with negative outcome—on the hypothesis that the brain operates via some sort of connectionist or deep learning net. A rigid, permanent, connection set is already, we might note, a hopeful idea. The brain is a fluid mass, constantly changing. It is trillions of particles, an enormous number of cells, in constant flux, transition, death, motion, transformation.

Lashley wondered if some kind of *field* was responsible for the remembering: "We have to consider that there are fields of force operating in the brain." So, he was considering a field impressing itself on the neural mass, perhaps an electromagnetic (EM) field. Sperry tested the EM-field hypothesis (cf. Gunther, Chapter 6 for a discussion [Gunther2012], and also [Gregory1987], p. 115). He used cats in form-recognition tests. His intent was to disrupt any operation of an EM field: "… transitory EM fields arising between nerve-cells …" He made tiny, crisscrossed cuts throughout the visual cortex, then riddled it with tantalum wires to short-circuit any fields, then implanted leaves of mica, all to stop transverse currents. The cats showed no loss in form discrimination. Sperry concluded that EM fields are not involved, and form perception, he speculated, depends on information traveling to small cortical regions via links "below the grey matter." That is, Sperry went right back to the concept that the wiring of the brain is fixed.

This is where it ended. But there is just one problem: Sperry did not make Lashley's findings "go away." The questions remain: Where is the memory? What is the memory?

HIGHLY SUPERIOR AUTOBIOGRAPHICAL MEMORY (HSAM)

From something that seems like a motor memory or procedural memory, let us turn to experience itself, to all those experiences of piano practice sessions. There have been indications noted for a long time that all experience is still accessible, that it is all somehow stored, but detailed, accurate, at-will access is just not the normal case. Oliver Sacks (*The Man Who Mistook His Wife for a Hat*), for example, describes "Martin A.," suffering from a form of brain damage, who could nevertheless quote verbatim *Grove's Dictionary of Music and Musicians*—any of its six thousand pages [Sacks1987]. He heard these quotes in his father's voice—memories from the long hours his father devoted to reading and sharing the history of music with his beloved, although handicapped son. Sacks's retardate "Twins" could, given a date in their lives anywhere after roughly the age of four, give its details in total—the weather, the political events of which they might have heard, personally related events—as though they were simply reviewing a vast panorama unfolding before their inward eye.

This kind of phenomenon has more recently been termed *highly superior autobiographical memory* (HSAM). Coming more into the public eye, several cases were featured on the Australian TV *Sixty Minutes* [Sixty2018]. The memory researcher, Eleanor McGuire, made an effort to at least note the phenomenon in her theory of the function of the hippocampus [RoyalInstitution2014] [McGuire2016]. Quoting an HSAM person's experience: "I am thirty-four years old. I can take a date between 1974 and today [2006, i.e., from 2 years-old], and tell you what day it falls on, what I was doing that day, and if anything of great importance occurred on that day; I describe that to you as well" [RoyalInstitution2014]. The same HSAM-person states that when a date flashes on the television, then "…I automatically go back to that day and remember where I was, what I was doing, what day it fell on and on and on …" [RoyalInstitution2014].

The difficulty for current memory theory, in truth, more of schizophrenia in the theory, is its presumption of *abstractionism* [Cf. Crowder1993]. All models presume that only *abstractions* of events are stored, or equivalently, "elements" of events, or just the "gist" of events. Mayes and Roberts are typical: "Only a tiny fraction of experienced episodes is put into long-term memory storage and, even with those that are, only a small proportion of the experienced episode is later retrievable" ([Mayes2001], p. 1398).

We've already seen the source of this in the propensity of current memory theory to break events into "features" that will be stored (and chemically "consolidated" in the brain over time) in various cortical areas. Note again the Moscovitch et al. [Moscovitch2016] assumptions, numbers three and four:

3. Encoding the event (or a fraction thereof): The hippocampus becomes an "index" to multiple aspects of the event stored in multiple areas of the brain.

4. Retrieving the event: The HPC (hippocampal complex) index is engaged, retrieving and reassembling the multiple cortically stored aspects of the event.

Note assumption 3—a fraction of the event (like stirring coffee for breakfast), or more precisely, "elements" (features) of the event stored in various cortical areas. As noted in Chapter 2 of this volume, there is no principled mechanism for the selection of this "fraction" or of the features. Elsewhere, one of the Moscovitch et al. authors, Lynn Nadel [UCICNLM2018], invokes Teyler & DiScenna [Teyler1986] as a paradigm for the "indexing" of Assumption 3, yet an *intrinsic* feature of this indexing model is throwing unreinforced or *insignificant* experience away! All of this flies in the face of HSAM, where everything seems to be "there" (or "encoded") down to what the person had for breakfast on any given date. It means for these HSAM folks either, (a) this fraction-selection mechanism must be entirely turned off, or (b) the HSAM "selection" method is mysteriously on steroids, selecting 100% of all possible "features" of any and all events experienced. Unfortunately for (b), the experienced event is not composed of "features." This is an abstraction imposed by the memory theorist.

We have been setting the stage here to question, perhaps better, dimming the lights on, any notion that the hypothesis that experience is stored in the brain has been proven—and *hypothesis* is all that it is—or that there is some ironclad model for how this is accomplished, even when the brain-storage thesis is granted: (a) a purely *static* framework of "features" to work within, and (b) discrete sampling via discrete "instants" as opposed to the true dynamics of events and the invariants defined over their flow. Many more problems around this concept (i.e., this hypothesis) could be explored, for example, the concept of "consolidation" of memories in the brain, chemically, over time (a process spanning a period always vague, unspecified— years? forever?), or the role of the hippocampus and its "index" as driving this consolidation (versus some other role). Yet the role of the hippocampus in some sort of spatial and temporal organization of the event being *currently* perceived has been repeatedly noted (e.g. [McGuire2016], [Buzsáki2017]). This may well be only to say that the hippocampal complex is critical in registering the spatial-temporal aspects of the ongoing event, along with all those invariants, say, over the ongoing coffee stirring, all which are being supported globally over the resonating brain. This brings in Fuster's comment in this context [Fuster1994], noting that what was discovered over the course of research was that, "... all 'storage' areas were simultaneously *processing* areas," which is to say, these areas may *only* be *processing* areas. Other issues involve the deeper implications re *actual* loss of experience as opposed to *apparent* loss in the literature on amnesia or aphasia (cf. [Robbins2009], [Robbins2020]). Of course, it goes without saying that this storage hypothesis, however, is taken as the very foundation, the bedrock of the computer or neural network model of mind, for it is obviously precisely within the computer's memory, direct-access storage device (DASD), or whatever, that "experience" is considered stored.

So let us turn to the model of memory retrieval when experience is taken within the Bergson-Gibson framework of mind, where the brain (with the body) is the leading edge, shall we say, of a temporally extended, four-dimensional being.

REDINTEGRATING EXPERIENCE: DIRECT MEMORY

Redintegration, it is always fun to observe, is a term that has been around since the inception of psychology. The "law of redintegration" was stated by Christian Wolff, a mathematics professor and disciple of Leibniz, in his *Psychologia Empirica* of 1732. In effect, Wolff's law stated that, "When a present perception forms a part of a past perception, the whole past perception tends to reinstate itself."

In everyday experience this phenomenon is encountered all the time. Thus, the sound of thunder may serve to redintegrate a childhood memory of the day one's house was struck by lightning. A person can be walking down a road in the summertime and notice a slight rustling or motion in the grass along the embankment. Immediately, they recall an experience in which a snake was encountered in a similar situation. Piaget's little daughter, Jacqueline, sees a piece of grass and immediately recalls playing with a grasshopper with her brother the previous day [Piaget1954]. Klein, in *A History of Scientific Psychology*, notes that these remembered experiences are "structured or organized events or clusters of patterned, integrated impressions," and that Wolff had in effect noted that subsequent to the establishment of such patterns, the pattern might be recalled by reinstatement of a part of the original pattern [Klein1970].

It is Gibson and ecological psychology that have been devoted to describing these "event patterns," that is, the invariance laws describing events. We could traverse the history of memory research and theory for the last near three hundred years since Wolff, noting the all-out avoidance of Gibson with respect to this memory subject, a history that extends throughout the "verbal learning" period with its focus on memory using only verbal materials (memorizing nonsense syllables, word lists, learning lists of pairs of words or "paired-associates," e.g., SPOON is paired with COFFEE), to the initial mathematical models (CHARM, [Eich1985]; TODAM [Murdock1982]; or Anderson et al. [Anderson1977]) which started representing words as vectors of 1s and 0s, for example, a vector for the stimulus word, "spoon" and a vector for the to-be-learned response-word, "coffee," and using matrix operations to simulate and explain retrieval given "spoon," remembering the response paired-word, "coffee", this in fact being already equivalent to a basic neural net conception.

There were, of course, misgivings about these vectors of 1s and 0s. What did these actually represent? Murdock noted that these were of entirely undefined psychological significance. One can increasingly improve memory performance by increasing N, the number of elements in the memory vector, but, he asked, "… what are these elements? If it takes 50, 100, 1000, or 10,000 elements to produce the necessary results, they are certainly not the cognitive features others have in mind" ([Murdock1982], p. 625).

Elsewhere, this history has been reviewed [Robbins2006] where the memory tasks edged ever closer to the ecological case, that is, toward concrete events with their dynamics and invariants and thus edging toward Gibson's principles, but have never gotten there.[1] The materials moved from employing the purely verbal (first, nonsense syllables, later, words), to vast discussion of the "concreteness effect" involving why "concrete" words [Paivio1994] were more effective for memory performance (e.g., "spoon" versus the more abstract "utensil") to imagery effects where subjects effectively imagined *events* for the two to-be-paired words, e.g., for SPOON-COFFEE, a "spoon" stirring the "coffee" [Piavio1971], to the concrete actions (mid-1980s) of "subject performed tasks" (SPTs), where subjects concretely performed (acted) long lists of concrete events, for example, "stir the coffee," "break the pencil," and "bend the wire," where memory performance (recalling the set of tasks) far exceeded the verbal case, and further, broke a few of the "laws" discovered

---

[1] [Robbins2006] treated several findings/areas of memory research via an ecological model: Cued recall, imagery effects, concreteness effects, priming, encoding specificity, interference effects, eye-witness phenomena, spreading activation, and several aspects of subject-performed tasks: (a) experimenter performed (EPT) versus SPT performance, (b) the role of objects, (c) the transitive-intransitive pattern, (d) verb-to-verb failure, and (e) EPT/SPT and reenactment.

by the verbal learning tradition [Zimmer2000], [Zimmmer2001a], [Zimmer2001b].

If we turn to the Gibson framework, the "law" for describing when a present experience will retrieve or redintegrate a past experience is fairly simple. It assumes a fundamental symmetry between perception and memory: the same invariance laws which determine the perception of an event also drive remembering. This implies a basic law for the fundamental operation of redintegration or direct retrieval:

1. An event E' will reconstruct a previous event E when E' is defined by the same invariance structure or by a sufficient subset of the same invariance structure.

In essence, when the same dynamic pattern for a present event (E'), supporting the same invariance structure, is evoked over the global state of the brain, the correspondent past experienced event (E) is reconstructed, or $E' \Rightarrow E$. We are essentially relying upon the same mechanism Gibson argued supported the direct specification of an event, and this is why it can be termed a model of direct retrieval or *direct memory*. We can take this in the context of our coffee stirring event. Let us restate coffee stirring's invariance structure for easy reference:

- a radial flow field defined over the swirling liquid
- an adiabatic invariant re the spoon, that is, a ratio of energy of oscillation to frequency of oscillation
- an inertial tensor defining the various momenta of the spoon
- acoustical invariants
- constant ratios relative to texture gradients and flows for the form and size constancy of the cup
- a ratio (termed "tau") related to our grasping of the cup
- and more …

To retrieve a past event of coffee stirring, E, we require a present event (E') with the same dynamic structure (or a subset). The brain, resonating to the structure of E', in essence is sending a resonant wave pattern through the 4-D extent of memory, through our 4-D experience. All events with similar structure are being retrieved—a sympathetic resonance. Now had we only one previous experience of coffee stirring, this is the only event retrieved, or equivalently, were something *unique* in this structure of E', for example, a fly moving by the cup slowly, like a heron, and E' holds this aspect as well. The uniqueness of E/E', structurally, or in a larger structure involving event context, allows for the retrieval of a specific event.

Given there are multiple experiences of coffee stirring, driving this resonant wave through the 4-D extent of being is equivalent to *dynamically defining* the *concept* of coffee stirring. In this operation, the variants (e.g., cup color, spoon form), so to speak, wash out, the invariants remain. In fact, as Cassanto and Lupyan [Cassanto2015] argued, *all* concepts are dynamic or in their terms *ad hoc*, and this mechanism, this dynamic definition, is the underlying reason why this *ad hoc* feature must be so. Numerous other theorists have envisioned something akin to this. The computer scientist, Gelernter [Gelernter1994] would conceive of a memory operation as taking an entire "stack" of memories or experiences, obtaining "focus" on one invariant aspect, which by this focus defined a certain level of abstraction. Goldinger [Goldinger1998] following Semon [Semon1909], and other "episodic" or "exemplar" models such as Hintzman's MINERVA 2 [Hintzman1986], would see an event activating "traces" left by all previous events (cf. also [Crowder1993]). The subset of partially redundant traces most strongly activated thereby defines, again, an abstraction, an invariance across the very concrete, experienced events.

But as far as remembering via E' as a cue and *locating E as precisely an event in our past*, this requires more than just the uniqueness of E. This is *explicit* memory. This is a problem of consciousness. It is another hard problem, in fact a problem that Chalmers and the consciousness field fail to recognize. This will require further discussion as we continue through the chapter.

MORE ON REDINTEGRATION

The more unique the event-invariance structure, the easier it is to retrieve the event. It is basically as if a series of wave fronts, w_i, were recorded upon a hologram, each with a unique frequency

(f_i) of reference wave. Each wave front (or image) can then be reconstructed uniquely by modulating the reconstructive wave to each differing frequency, f_i. The unique, coherent frequency avoids "interference" of the events [McGeoch1942], that is, *multiple* wavefronts/events (or multiple words given verbal stimuli) being simultaneously retrieved by a cue (like the fuzzed image of a cup/cube via the noncoherent f_1/f_2 discussed in Chapter 3, Figure 3.2). This implies a second law for sets of events:

2. Given a set of events, $E1$, $E2...En$, the more unique the invariance structure defining each event, the greater the probability that each will be reconstructed by events $E1'$, $E2'...$ En', with respectively the same invariance structures or a subset thereof.

Thus, suppose a series of perceived events: (1) a man stirring coffee, (2) a baseball hurtling by one's head, and (3) a boot crushing a can. Each has a unique invariance structure. To create the reconstructive wave for these, that is, to evoke over the brain the needed modulatory/dynamical pattern, we might use as a "cue" respectively: (1) a stirring spoon, (2) an abstract rendering of an approaching object capturing the composite tau value of the original event [Craig2000], and, (3) an abstract rendering of one form descending upon and obscuring another [Verbrugge1977]. But these events are *multimodal*, and the four-dimensional extent of experience is multimodal. There are auditory invariants as well defined over the events. Our cues could become respectively: (1) the swishing or clinking sound of stirring, (2) the "whoosh" of the passing baseball with frequency values capturing its inherent doppler effect (coordinate with the approaching baseball's "looming"/expanding flow field), (3) the crinkling sound of collapse of a tin structure. And in the dynamics of the haptic component of the event, we could cue our stirring event by wielding a "tensor object" that captures the inertia tensor (invariant), I_{ij}, specific to spoon-stirring [Turvey1995].

One can imagine then a quite "fearsome" task in the verbal-learning tradition's paired-associate paradigm as far as verbal learning experiments is concerned. A list for experimental subjects to learn would look as follows:

SPOON-COFFEE
SPOON-BATTER
SPOON-OATMEAL

SPOON-BUTTER
SPOON-CORNFLAKES
SPOON-PEASOUP
SPOON-CATAPULT
SPOON-CHEESE
SPOON-TEETER TOTTER
And so on …

It is fearsome from a verbal learning perspective since the same stimulus (cue) word appears constantly, thus providing absolutely no clue to which response word is intended. It is the extreme case of the *A-B*, *A-C* list paradigm where two lists of words are learned successively, where the pairs share the same stimulus word, for example:

List 1 (*A-B*)	List 2 (*A-C*)
SPOON-CUP	SPOON-PLATE
BOAT-LAMP	BOAT-TABLE
KNIFE-SOAP	KNIFE-MEAT

And so on …

It is again fearsome since, in the purely verbal case, (a) the stimulus (cue) word retrieves *both* responses, "spoon" ⇒ "cup/plate" (i.e., "interference") and, (b) the subject could have no clue what the appropriate response word is (other than keeping the *context* straight—"These words are list 2 versus list 1" —implying remembering which pairs were in which list). But assume that the subject concretely acts out, while blindfolded, each event of the "absurd" list—stirring the coffee, stirring the cake batter, scooping/lifting the oatmeal, the cornflakes, cutting the cheese. To effectively cue the remembering of what paired object was involved, as already noted, the dynamics of each cue-event must be unique, and the structure of invariance laws of an event in effect implies a structure of constraints. This principle, we earlier alluded to, was already being illuminated as a few researchers switched their memory tasks (in the mid-1980s) from the verbal materials to what were in effect ecological tasks, the subject-performed tasks (SPTs), where real dynamical forces with invariants—concrete events—were involved, with subjects performing long lists of actions, for example, "break the pencil," "lift the cup," "stir the coffee," and tasked to remember the list. Memory performance was far greater; the supposed "laws" derived via verbal learning experiments were altered.[2]

PARAMETRIC VARIATION OF MEMORY CUES IN CONCRETE EVENTS

The constraints in these ecological events may be "parametrically" varied, where increasing fidelity to the original structure of constraints of a given event corresponds to a finer tuning of the reconstructive wave. The (e.g., blindfolded) subject may wield a stirrer in a circular motion within a liquid. The resistance of the liquid (a parameter value) may be appropriate to a thin liquid such as coffee or to a thicker medium such as the batter. The circular motion (a parameter value) may be appropriate to the spatial constraint defined by a cup or to the larger amplitude allowed by a bowl. The periodic (back and forth) motion of the hand may conform to the original "adiabatic" invariance (the ratio of frequency/energy) within the event or may diverge. We can predict that with sufficiently precise transformations and constraints on the motion of the spoon (either visual, auditory, kinesthetic, or combined), the entire list can be reconstructed, that is, each event and associated response word. Each appropriately constrained cue-event corresponds to a precise modulation (or constraint) of the reconstructive wave defined over the brain.

The obvious inverse is that, as the parameter values diverge from the original event, cueing/recall performance and/or recognition performance will increasingly degrade. Recognition tests are one method of testing these manipulations, employing familiarity ratings. In this case, we would present, as recognition cues, the original events transformed on various dimensions. Experimental subjects rate the re-presented event with a value reflecting how sure they are that they have previously seen it (how familiar). The familiarity value should steadily decrease as the parameters vary away from their original values.

---

[2] See [Vicente1998], [Vicente2000] for an ecological memory model taken at a higher level of event abstraction, for example, chess or baseball games (realizing this is in effect discussing constraint-attunement on the same basis as the event-invariance structures discussed here).

These kinds of experiments are implicit in the literature. They are just not realized as "fits" in the theoretical structure we have just described. An example is found in a demonstration by Jenkins, Wald, and Pittenger [Jenkins1978]. Capitalizing on the notion of the optical flow field, they showed subjects a series of slides that had been taken at fixed intervals as a cameraman walked across the university campus mall (Figure 4.1). Some slides, however, were purposely left out. Later, when subjects were shown various slides again and asked if they had seen the slide shown, they rejected easily any slide taken from a different perspective and which therefore did not share the same flow field invariant defined across the series. Slides not originally seen, but which fit the series with its flow field were accepted as "having been seen" with high probability. But Jenkins et al. had created a "gap" in the original set shown to the subjects by leaving out a series of six continuous slides. Thus, a portion of the transformation of the flow field was not specified. Subjects were quite easily able to identify these slides as "not seen." In this case, we are in effect varying parametric values defining a flow field. We are seeing that redintegration, and therefore the brain's retrieval dynamics, is indeed sensitive to this form of parameter manipulation. Other manipulations are possible, for example the slant of the gradient [Domini2002], the smoothness of the flow, the velocity of the flow, and so on.

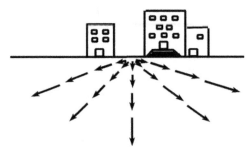

FIGURE 4.1. A flow field is created while moving along the campus mall.

We can consider, for example, presenting a simple static event of a field with a schematic tree (Figure 4.2). The trees in the figure have been grown with the precise mathematics defining real tree growth [Bingham1993]. The number of terminal branches (N) is a set function of height (H), $N=k(H)2$, while the diameter (D) of trunk or branches at any point is a function of remaining length to the tip, $D=k(H)$. Suppose that one of the recognition items in our experiment is the leftmost or youngish tree in Figure 4.2. In the recognition test phase, the parametric values defining the structure of the trees can be varied increasingly from the original value. For example, we could re-represent various trees at different stages of growth as defined by the math parameters and ask if this is recognized as part of the original set of items. Familiarity ratings should drop the further we move along the age dimension. More dynamically, a time-accelerated view of the tree's growth under certain parametric values can be used as the stimulus. For a dynamic event such as an approaching rugby ball, the value defined over the velocity flow of the approaching ball ("time to contact"), which specifies the immanence of the object hitting us, can be varied increasingly from the original value [Craig2000].

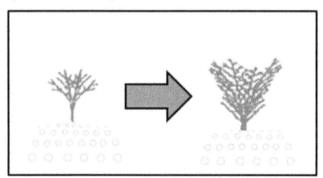

FIGURE 4.2. Two generated trees, one young, one old. (Adapted from Bingham, 1993.)

Even what, for us, is a very slow event, namely, the aging of the facial profile, has its invariance structure. Pittenger and Shaw [Pittenger1975] showed that this event can be described and generated via a strain transformation on a cardioid (Figure 3.5). Their original experiments can be recast in this redintegration test framework. Originally, the subjects looked at many pairs of faces, each face being generated by this law with varying values of the strain transformation, and judged each time which of the pair was the older. Changing this to a memory task, a face of a certain age can be included in a set of various items successively presented to a subject. On the recognition task, a face transformed by a certain parametric aging value is now presented. Familiarity values will be a function of the closeness to the transformation or strain value.

The aging transformation works for animal faces too, even for Volkswagens—it can generate increasingly aging "Beetles!" So, we can have many different kinds of items in this test that eventually get aged (or un-aged) in a recognition phase.

All these are memory experiments that are either waiting to be done or implicitly have been done.[3] They demonstrate that memory and its supporting brain dynamics are extremely sensitive to the invariance structure of events and the actual "parameter" values of the transformations involved. The importance of this is that *any theory that claims to be a model of memory must be able to support this dynamic structure in events.* But as we shall see, the major claimants to the theory of memory, namely neural networks, or connectionist networks, are utterly incapable of this—they can *approximate* this sensitivity to these laws in a vague way, they cannot truly capture it.

ANALOGICAL REMINDING

Redintegration is inherently *analogical*. When Piaget's daughter, Jacqueline, sees a piece of grass and there is the reminding of playing with a grasshopper yesterday, we are dealing with an analogy—blade of grass to grasshopper. The critical nature of this fact cannot be overemphasized, and no work has delved into this key area more than Hofstadter and Sander in their book, *Surfaces and Essences* [Hofstadter2013].

"Analogy is the fuel and fire of thinking," stated Hofstadter and Sander. By this, they meant precisely that "... without concepts there can be no thought, and without analogies there can be no concepts (p. 3)." To elaborate further, they envisioned that, "... each concept in our mind owes its existence to a long succession of analogies made unconsciously over many years, initially giving birth to the concept and continuing to enrich it over the course of our lifetime (p. 3)." Hofstadter and Sander see the scene of our thought as one of constant, voluminous, continual analogy. Analogy is the fundamental operation of thought. This fundamental operation is itself subserved

---

[3.] The many memory experiments of Jennifer Freyd and colleagues, often around "representational momentum," also fall into this category of parametric variation of dynamic memory cues [Freyd1987], [Freyd1984], [Freyd1990].

by an operation of memory (yes, we have already seen this operation, namely, redintegration), constantly dragging our past experience back into the present, ever placing the present event, via the context of the past, in an analogical light.

An analogy that serves a salient purpose for Hofstadter and Sander features an event where the main author (DH) is visiting the North Rim of the Grand Canyon with his wife and young son, Danny. In the midst of this scene of great majesty and splendor, Danny bends down, intent on watching some insects in the sand. Many years later, when DH is visiting the great temple of Karnack, his friend, Dick, suddenly seems more focused on bending down and peering at a small bottle cap on the ground (bottle caps being objects for which he has an obsession for collection). The former scene at the Grand Canyon comes rushing back to DH in memory. We have, in other words, an *analogical reminding*.

Analogical reminding was a term employed by AI theorist, Eric Dietrich [Dietrich2000], as he contemplated the problem of analogy. Walking in an alley, he spots a jumble of garbage cans. Instantly he sees "Garbagehenge." In essence, the static structure of the garbage cans has driven recall of past experiences of Stonehenge, whether of visits or pictures, and in a meshing of these, the lowly garbage cans are now exalted in what we can call an "analogic" event. In this little example, Dietrich was struck by the fact that at the basis of analogy is an apparently instantaneous operation of memory retrieval that, of itself, is inherently analogic. In fact, he argued, it is this *analogic operation itself that is defining the features of the current event* (see also [Indurkyha1999], [Robbins1976]). Currently, as we've already noted, the inverse is the case, that is, it is held that the already-defined features of a current event drive or define an analogy to a past event with the same features (as for example, [Gentner1983], Doumas, Hummel and Sandhofer, 2008 [Doumas2008], and for a critique of the predefined features-aspect of these models, see Chalmers, French and Hofstadter [Chalmers1992], and also [French1999]). Applying Dietrich's view in the Danny case, the retrieval of the Danny/Grand Canyon event and its superposition upon the Dick/Karnack event is itself defining the features of the analogy—"attention on the trivial in the context of a grand geological/architectural feature." As an AI theorist, Dietrich acknowledged that he had no idea how such an operation could be implemented.

In the Danny analogy, the dilemma for Hofstadter and Sander thus begins. They now begin the search for the mechanism of analogical reminding, that is, the basis for this fundamental operation of memory. The authors ponder how the original event was *encoded* such that the analogy with the future event is perceived (or the reminding caused). The notion of "encoding" (Danny and the insects at the Grand Canyon) they note (p. 161 and ff.), is really how the event is being *conceptualized*. The term should not be taken as the "encoding" of the event in the brain as though storing a literal string—"Danny looking at trivial insects in the presence of a major geologic feature." It is actually the *category*, namely, "trivial sideshow more fascinating than the main event." Thus, they stress, "… it is not the sequence of English words that we are talking about, but the abstract idea that it denotes ([Hofstadter2013], p. 171)."

The problem is twofold: First, there is far more to the event than just this conceptual structure (if indeed this is encoded at the time, as opposed to Dietrich's dynamic creation via the analogy or, equivalently, Cassanto and Lupyan's constantly ad hoc concepts), and this "far more" is what the experience available to retrieval is. The conceptual structure is defined within the event, but what is the conceptual structure, in and of itself, without this event (and/or others) over which this structure (abstraction) can be defined? What is an abstract idea if not a pattern defined over or within concrete experience(s)? What is "a pure abstraction" (or "conceptualization")? "Bending," for example, involves something being bent, whether an elbow, a wire for a hanger, a branch for a bow, or Danny's body bending down. "Bending" may be a concept, but as an abstraction, it does not exist without the concrete events of something being bent. Intrinsic to these bending events, like the coffee stirring, are very concrete forces, invariance laws, and visual transformations.

Second, this "far more" to the event can support many other "encodings." A guide at Karnack, extolling its grandiosity, could have served as well to drive the retrieval of the Grand Canyon experience (perhaps also with a Grand Canyon-extolling guide) to include Danny's presence, or a sudden breeze stirring the desert sands at Karnack could retrieve the Grand Canyon experience which also could have had a sudden breeze, or perhaps a hawk soaring over Karnack serves for retrieval, or the conversational mention of buying hotdogs by the souvenir stand, or walking up a steep slope to the Karnack temple, or Danny's flapping shirt-tail, and so on. Is there then a limit to the possible encodings?

Recognizing this latter underlying problem, they ponder, reluctantly—shades of HSAM—the possibility of *all experience, in detail*, being stored in the brain. Their hypothetical model was of videos stored on a DVD (as though some form of instantaneous molecular transcription mechanism has been discovered) where experience would be captured in its entirety in the neuronal structure of the brain. For Danny and the Grand Canyon, the entire (multimodal) event would have been "filmed" or recorded in DH's brain while he was experiencing it. Thus, some twenty years later, "… when he observed Dick stooping to pick up a bottle cap, this specific film had been reactivated in his brain by a mental search algorithm running through all filmed scenes in his memory" (p. 172).

The impossibility of this vast storage (*in the brain*) of "event videos" is obvious, if only first because modern neuroscience can find nothing approaching an area in the brain that stores the replication of whole events, nor, for that matter, has anything like a molecular transcription process, acting in real time, been discovered. The authors further disparage the existence of anything like "pure visual resemblances" that could be used for matching events. Rather, they argue that the lesson here is that we do not store in our memory a collection of "objective" events through which we search, seeking perceptual resemblances, whenever new events happen to us, but, "… rather the events that befall us get encoded—that is, *perceived, distilled and stored*—in terms of prior concepts that we have acquired" (p. 172, emphasis added).

In other words, they are just succumbing to the "abstractionist" position here—that only the "gist" of events, or "features" or "elements" are "encoded," that is, stored. But the authors are deeply perplexed on the subject of the very principles (for there are none) which govern the encoding of a concept for an event, that is, just when, or yes, *if*, these abstract "essences" are indeed recorded as the "surface" of the event unrolls. Was the concept, "trivial sideshow …" really encoded as DH watched Danny long ago? They have already voted, as we have seen in the previous text, that this encoding must exist, yet they are utterly uncomfortable with this very notion. For one thing, a reliance on such "encodings" of a previous event to explain a subsequent analogy seems very fortuitous. Second, as already noted, there is the question of how many concept encodings an event might have, yes, just as the many possible "functions" of a pencil, for the event comprising the Grand Canyon visit is very doubtfully limited to supporting just one analogy. Third, requiring previous encodings would appear to limit the emergence of analogies to precisely the set related to any previous encodings, yet the very "essence" of their book describes the inexhaustible, ubiquitous, continuous, creative, massive emergence of analogies via past experience.

In other words, analogical reminding [Robbins2017] requires the operation of redintegration, where the invariance structure of the current event, E', sends its "wave" through our 4-D experience. Intrinsically, then, concepts become dynamic, ad hoc, continually redefined. But such a conception, as the struggles of Hofstadter and Sander display, is unavailable without the concept of such a 4-D memory.[4]

CONNECTIONIST NEURAL NETS AND THEIR FAILURE IN ANALOGY

While the GPT technology might be thought to have totally superseded the connectionist network approach in the AI world, the two are "apples and oranges." Neural networks obviously underlie both frameworks, but the connectionist approach was directly aimed at providing an associationist model of knowledge; the GPTs do so more indirectly. Connectionist nets were proposed— as a few of an enormous number of examples—as a model for the paired-associate learning of the verbal learning tradition, even as an

---

[4.] While we have tended to refer here to the *brain* as the reconstructive wave, note that given the principle of virtual action, thus the involvement of motor areas and their vast connections within the body, the *whole body* should be seen as involved in this wave, and this includes the *heart*. We note this due to the phenomena involved with heart transplants, where the memories/experience of the previous owner of the heart are now experienced by the recipient (cf. [Pearsall1999]), This is just one of numerous cases where, (a) clearly, the hypothesis that experience is stored in the brain is contradicted (cf. Sheldrake for others [Sheldrake2012]), and (b) where it seems the whole body need be taken as supporting the reconstructive wave.

ecological model for our knowledge of events, for example, given a bird sitting on a branch, we know it will fly away, that is, these two features of the event—bird-sitting, then-flying—are associated via a connectionist net. This latter "ecological" feature was the claim of Rogers and McClelland [Rogers2004]. The GPT, as we will look at more closely later (Chapter 7), with its encoder component, starts at its base with a corpus of words in text, where each word is a vector whose component values measure its distance/relation from other words. Given the "attention" component, also a part of the architecture, the GPT can determine that in a sentence like, "He swam across the river to get to the other bank," that "bank" must be clustered here, so to speak, with "swam" and "river," meaning we are likely dealing with the bank of a river in this case, not a bank for holding your deposits. Of course, this transformer methodology can do much more, including tell us, from its "knowledge" based in the text corpus, how to build a mousetrap from a box, pencil, razorblade, and so on, but this approach is oriented to and based in something rather (but not entirely) different than the connectionist architecture, an architecture that was already purporting to be an association model for understanding ecological events, or doing paired-associate learning, or explaining memory loss phenomena arising from damage to the hippocampus [McClelland1995]. The point here is that the connectionist approach has not actually been replaced, in fact, some GPT-enthusiast theorists (e.g., [Piantadosi2023], see Chapter 7) think it *still* holds the key to semantics beyond the semantics that GPTs are held to accomplish. It still needs to be critically examined.

So, the last point in the previous section looked at the question of whether there is a limited, finite set of encodings that an event might have (like looking at the Grand Canyon). To reinforce this, consider the Hofstadter and Sander discussion ([Hofstadter2013], pp. 189–190) of all the "categories" a glass can participate in. Categories, they argue (pp. 434–436) *are simply analogies*, discussed confusedly as though two separate things in the cognitive science literature, for example, "Categories let people treat things as if they were familiar" [Spaulding1996], versus, "Analogy is what allows us to see the novel as familiar" [Gick1983]. Just some of the things in the "glass" categories ([Hofstadter2013], pp. 189–190), with our own mixed in, are listed here:

▪ artifact, industrial product, consumer article, fragile object, item of dishware, drinking glass, water glass, boat bailer, transparent object, trap for flies (dome), recyclable object, projectile

▪ pencil holder, toothbrush holder, flower vase, piece of freight, piece of merchandize, item for sale, unsold object, unsellable object, temporary fishbowl, dust-gatherer, glass for cold drinks, home for tadpoles

It is not necessary to engage in argument here as to whether this "categories as analogies" argument is definitive—we are focused on the problem with finite encodings placed upon events versus the inexhaustiveness of analogical remindings. This exercise is reminiscent of French's [French1990] considerations on the number of events that an object, for example, a credit card, can participate in. In the context of the Turing Test, French proposed various tests for any computer attempting to masquerade as a human, where obtaining a passing grade relied totally on having the requisite concrete experience. One test was a rating game, with statements such as:

▪ Rate purses as weapons.

▪ Rate jackets as blankets.

▪ Rate socks as flyswatters.

The computer's ratings would be compared to human rating norms. French argued that there is no way a computer can pass such a test without the requisite concrete experience. The problem equally holds for evaluations of statements such as:

▪ A credit card is like a key.

▪ A credit card is like a fan.

The list is endless. French states, "… no a priori property list for "credit card," short of all of our life experience could accommodate all possible utterances of the form, 'A credit card is like X (p. 94)." Without the experience, he noted, one incurs the necessity of either preprogramming or training-up the association weights of all possible pairs of objects.

This "training-up of all possible pairs," it is worth noting, is exactly the thicket that connectionist models of memory walk into when we consider our powers of analogy. The network of Rogers and McClelland [Rogers2004] [Rogers2008], for example, can be trained, given the input pair "SPOON CAN," to eventually respond with an output set such as STIR, SCOOP, LADLE, HOLD WATER. This finite set is the networks specifically trained "knowledge" or encodings on what a spoon "can do." Yet, when we imagine little brother launching a pea with his spoon at big sister across the table, we can easily see this addition to French's list:

- A spoon is like a catapult.

This (catapulting) is now something else that the spoon "can do," but the network, not having been trained on this response, would be incapable of recognizing it as a legitimate possibility [Robbins2008]. Yet this derives naturally, as an analogy, from our experience of catapulting, whether by reading about Roman catapulting or watching "Pumpkin Chunkin" contests with pumpkins being flung by trebuchets.[5] For French, therefore, the number of "features" a credit card (or a spoon) can exhibit is inexhaustible. This is to say, equivalently, that we are unable to specify a limit to the number of concepts "encodings."

This capability—seeing the spoon as a catapult—is in effect *the projection of a possible component within an event invariance structure*. A spoon is being projected within the event of catapulting. Does it support the invariance structure (like a trebuchet arm does)? Does the component (say, an orange peel) support the structure,

---

[5.] One connectionist response is that there are better architectures that solve this problem. The vector symbolic architecture (VSA) [Gayler2003], offered as one example, is an effort to implement a more biologically plausible, more efficient architecture without using the troublesome backpropagation method of Doumas et al. [Doumas2008] or Rogers and McClelland [Rogers2008]. Nevertheless, the fundamental view of analogy is exactly the same, that is, analogy depends on a systematic substitution of components of compositional structures. In VSA, the systematicity and compositionality are considered the outcome of two operations, namely, binding and bundling, where binding associates "fillers" (spoon, coffee) with "roles" (stirrer, stirred), and where bundling combines these role/filler bindings to produce larger structures. Therefore, the very elements of the composition, that is, the features of objects, must be predefined (encoded), the values set, the relations fixed—all for the sake of allowing a syntactic process to unroll.

for example, the force, rigidity, width required for coffee stirring? Events are inexhaustible, thus the "features" as well.

CONNECTIONIST ANALOGY: MORE DYNAMIC PROBLEMS

Discovery of relations by analogy (DORA) [Doumas2008] is a connectionist model devoted to explaining the dynamics of analogy formation. DORA employs a *comparison* process across what it calls "events," or better, partial fragments of events (Figure 4.3). As it *assumes* there is a feature-fragmentation accomplished (somehow) by perception, let us consider this further for a moment. At this level, DORA envisions comparing object units, finding common features, and creating single place predicates. Thus, we might bring *spoon* and *spatula* into memory and find a common "feature," namely, "stirrer." This stimulates the birth of the explicit property, "stirrer," and this can be bound to each object, such that we have *stirrer* (spoon), or *stirrer* (spatula). *Every* feature in common between the two objects, according to DORA, is also linked to this relation (*stirrer*).

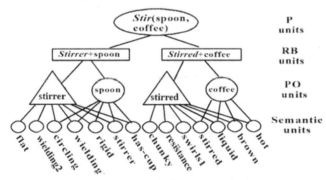

FIGURE 4.3. A representation of stirring in the feature framework of DORA. (P = Proposition units, RB = Role Binding units, PO = Predicate and Object units [which code for individual predicate and attributes and objects]). (Adapted from [Doumas2008].)

This is the first problem: If "stirrer" is a "feature" of these two objects, how can this "feature" possibly be other than a full stirring event with its invariance structure? Just what is a "partial" stirring event such that we only have a "stirrer" as a feature? Now, to find these two "features" of the two objects the "same," we must be implicitly *comparing* the two *events*. But the two events—spatula stirring of a cake batter and coffee stirring—have similar but different radial flow fields, different values of wielding, periodicities, acoustics, and so on. How does the model (with only its abstract symbols and tokens) determine these two "features" are the "same?" How does it achieve this without an actual physical comparison?

Consider briefly the nature of comparing the spoon's stirring versus the spatula's, expanding the already noted concept of the inertial tensor a bit. A rigid object's moments of mass distribution constitute potentially relevant mechanical invariants since they specify the dynamics of the object. The object's mass (m) is the zeroth moment, while the first (static) moment is mass (m) times the distance (d) between the point of rotation and object's center of mass. The second moment is conceived as the object's resistance against angular acceleration. In three dimensions, this moment is a 3×3 matrix (the inertia tensor). The diagonal elements I_1, I_2, I_3, are eigenvalues and represent the object's resistance to angular acceleration with respect to a coordinate system of three principal axes (cf. Kingma et. al, [Kingma2004]). There will be an inertia tensor (invariant), I_{ij}, specific to spoon stirring. Over the periodic motion of the stirring spoon, there is likewise a haptic flow field defined, and within this, there is an adiabatic invariant—a constant ratio of the energy of oscillation to the frequency of oscillation [Kugler 1987]:

$$\frac{\text{Energy of oscillation}}{\text{Frequency of oscillation}} = k$$

Again, what exactly would it mean to "compare" these two "wieldings" and periodic motions? How would the two events be synced in time-synchrony for the comparison? And the really basic problem: how can a connectionist architecture possibly represent these dynamics with its invariance over time?

A second problem: Figure 4.3 shows the dilemma, for there we attached features to the spoon such as wielding-1 (i.e., some arbitrary "value" for wielding the spoon), circling, rigid, and so on. Are these part of the object, or part of the event in which the object participated? In reality, they are part of *stirring*. If they are part of the spoon, how are they the "same" as the spatula's?

The origin of these dilemmas is the fundamental "simplifying assumption" of connectionism whereby the ecological world is viewed as *objects* and *properties*, with properties (features) attached to objects. But the simplifying assumption is simply assuming the problem away: it is *the definition of the features* that is the problem. Objects are simply special cases of an event. In turn, an object–event is always itself nested in some event—the spoon at rest is an event, or the spoon is stirring, it is scooping, it is cutting cheese, it is catapulting a pea. Its "properties" emerge over or during the specific event. It is events that are primary. And events have a structure over time that is resolutely ignored only with the consequences previously stated. The cheese-cutting spoon, the soup-stirring spoon, the pea-catapulting spoon—all show different properties that are not "directly apparent" in the resting spoon. "Properties" of objects are always functions of events. But the *abstraction* of these properties, the definition of the abstract features, this is a function of analogy, for the transformation that is analogy is defining the features. For this, we require a redintegrative device embedded in the indivisible transformation of the external field, which is to say, within 4-D experience.

GPT-4, we know, using its word-vector database, has scored in the upper one percent tile on the Torrance Test of Creative Thinking, performing impressively on questions like thinking up all the possible uses of a basketball [Guzik2023]. This appears closely in accord with French's rating tests or coming up with possible uses of a pencil. This is an amazing approximation given it is based solely on the mass of linguistic strings dwelling on the internet. It obviously has nothing to do with our experience of the invariance structure of events.

EXPLICIT MEMORY: CONSCIOUS LOCALIZATION IN THE PAST

Redintegrating an event, where $E' \Rightarrow E$, is a form of *implicit* memory, an unconscious—not at the level of awareness—operation. It is not equivalent to "locating the event as an event in our past," that is, a conscious localization of the event as something we have previously experienced. We see and hear the wind chimes tinkling on the porch and the event returns to where we remember buying the chimes in a gift shop for our wife. This is an *explicit* memory.

Although the implicit-explicit distinction is ubiquitous in the memory literature, there is no understanding in current theory as to why this is so, particularly why explicit memory requires consciousness or what this means. Remember, all current cognitive theories have no role for consciousness. *Implicit* memory as a concept arose with the discovery that amnesiacs can exhibit a form of learning (remembering)—a learning of which they themselves are not explicitly aware. The original discovery of this was with the case of a famous amnesic, H.M.

In 1956, William Scoville removed the medial portions of both H.M.'s temporal lobes in an attempt to control an epileptic condition. After this operation, the nature of his conscious experience changed markedly. He displayed severe amnestic syndrome, appearing unable to retain any further long-term memories. H.M. could not remember events from one day to the next, in fact for far briefer intervals. He could retain a string of three digits, for example, 8-2-4, by means of an elaborate mnemonic/rehearsal scheme for perhaps fifteen minutes. Yet five minutes after he had stopped and explained the scheme to experimenters, the number was forgotten. Remarkably, his same examiners would go into his room day after day yet he could not remember ever having seen them. It was Milner who discovered that H.M. could learn perceptual or motor skills such as mirror tracing. But, although steadily improving from session to session, he insisted upon entering each practice session, seeing the mirror-tracing apparatus and such, that he had never done this before. For H.M., by his own description, each day is "a day unto itself," without history.

Weiskrantz [Weiskrantz1997] noted the many things subsequently discovered that amnesiacs could learn: paired associates, artificial grammar, answers to anomalous sentences ("The haystack was important because the cloth ripped." Answer: parachute). There appears to be no limit; no limit, that is, as long as the task does not require the amnestic to place an event in the past (amnesiacs are terrible on the *A-B*, *A-C* paired-associate task noted previously, where one must keep straight what word-pairs were in List 1 versus List 2). Weiskrantz expressed these amnestic-capable tasks as tasks which do not depend on a "joint product" relating present to past (current event × stored event). The amnestic simply cannot perform this product, this fundamental comparison—present × past.

Clive Wearing lives perhaps in an even shorter temporal frame. Clive is another individual, similar to H.M., with extensive bilateral damage to the hippocampus, amygdalae, and so on. Clive, as H.M., is unable to remember previous events. He feels constantly that he "has just woken up," and keeps a diary in which this feeling is repeatedly recorded at periods of hours, even minutes. He makes statements such as, "Suddenly I can see in color," "I've been blind, deaf, and dumb for so long," and "Today is the first time I've actually been conscious of anything at all." To Wearing, the environment is in a constant state of flux. A candy bar, even though held in his hand, but covered and uncovered by the experimenter, appeared to be constantly new.

We have seen the mechanics of redintegration. Now we see that this mechanics must include some other "tones" as it were, beyond the fundamental, to explain direct recall of events as explicitly past events, that is, to localize these events as events in one's past. H.M. approaches for the nth time the mirror tracing apparatus. The same flow field is specified as he moves to the device, moving across the same texture gradients specifying the same floor and the same table surface, the same optico-structural transformation of the device as he moves toward it, the same action being carried out. Everything required for redintegration appears present—the same event pattern, the same perceptual information. The redintegrative dynamics should be perfectly supported. The phenomenon of priming, which amnesiacs still demonstrate to have, indicates that it is. But there is no explicit memory of ever having done, seen or experienced this mirror-tracing thing before. For a theory of consciousness, the problem is startling, with strange symmetries to the problem of perception. We have all the neural dynamics supporting perception and redintegration operating, yet just as in those neural/computer architectures that Chalmers saw as lacking "qualia," there is here also a mysterious ingredient missing—the ingredient that makes the past event *consciously* remembered.

The critical difficulty for consciousness is why, even in a fully functional "explicit memory system" (as current theory simply *labels* it) in which all retrieval mechanisms are fully operative, be they connectionist neural nets with their firing patterns, or the operations of cognitive symbolic programs—why should these operations result in a consciously experienced remembrance as opposed to just another "implicit" operation. This is the question: Just what is required for a past event to be consciously experienced as an event localized in one's past? What must such a system involve? As one might already intuit, this question is verging on the very nature of the *symbolic* in the framework of human consciousness.

CHILDHOOD AMNESIA, THE EXPLICIT, AND PIAGET

The true complexity of this past × present "product" is in general vastly underestimated, if discussed at all. The ability does not develop overnight. The literature of childhood amnesia sees the child requiring at minimum two years to achieve explicit memory. But this literature rather underestimates, in fact vacillates on, what is involved. Thus, there was a period in which there was an extensive effort to show that children evidence "access consciousness" [Block1995] considerably earlier than this two-year figure. Access consciousness, in its definition as "the ability to use previously stored representations to direct thought" is certainly subsumed in the developing ability to establish the past × present product. Rakison, in a review of these attempts, has shown that in every case, this (earlier development) thesis can be rejected [Rakison2007]. The two-year figure holds; the dynamic trajectory is developing a complex ability; it requires a certain time to unfold. The robotics theorist should be asking why so long a trajectory and what is being accomplished?

When the explicit is tied to "access consciousness," it is a somewhat common conception that it is language that gives rise to access consciousness (e.g., [Rakison2007]) and by implication, explicit consciousness of a past event. The developmental trajectory supporting the explicit about to be reviewed nowhere relies on language; it is language that relies on this trajectory, a trajectory which itself results in the ability to support the symbolic—the symbolic itself obviously being the foundational requirement.

So, it is generally accepted, as already noted, that it requires roughly two years for the child to develop explicit memory. In fact, this period coincides with something termed childhood amnesia, namely, the inability to remember the early events of childhood. Something is occurring in the brain, as a dynamic self-organizing system, over these two years which eventually allows the brain to achieve a complex dynamic state. This does not mean, by the way, that experience is not "stored" during these two years. It means that the brain requires two years of dynamic organizing to achieve this ability to perform a "past × present" product. In other words, infantile or childhood amnesia seems symmetric with lesion-induced amnesia in later stages of life where the dynamic neural organization that has occurred over our first two years to relieve the first state (infantile amnesia) and bring about explicit memory may be the same neural organization that has been damaged or lesioned, bringing about (or bringing back in a way) the second case, amnesia.

For infantile amnesia, in their book and excellent review of the subject, Howe and Courage [Howe1993] rejected explanations that relied on retrieval failures (e.g., repression, mismatches between initial encoding and later retrieval contexts) or storage failures (perceptual or neurological immaturity, inadequate "encoding"). To account for this discontinuity and the sudden growth of memories after the age of two, they evoke the emergence of the "cognitive self" around this age. Taking a page from connectionist theory, they argue that once the self develops its "features" (as though the self is a "vector" of features), these features can be incorporated into the memory "trace" along with features of the (external) event in which the child is participating. As more features of the self are accumulated, the more the probability of sampling a set of these features and including them in the memory "trace."

Howe and Courage are perhaps not so committed to the "feature" theory as they are to the general proposition that autobiographical memories can be accounted for by general properties of contemporary memory models, where the self takes its place "as a system of knowledge that organizes memories like any other knowledge structure" ([Howe1993], p. 513). The "features" of this self are not at all articulated, but noted are the interpersonal self, the ecological self (fundamentally the infant's response to

Gibson's invariants), and the conceptual self. The conceptual self "enables infants to take themselves as objects of thought" (p. 507). Here parent–child interactions are discussed, fostering the child's awareness of himself as an object of attention.

Howe and Courage contain one small reference to Piaget and his profound analysis of the dynamic developmental trajectory necessary to support this "self as an object of thought," that is, the very meaning of this phrase ("self as object") which comes from Piaget's analysis, is unmentioned. But this downgrading of Piaget has become the norm, so the norm that perhaps some are wondering why there should be any focus on Piaget. To a certain extent, it is fair to say that he has been written off by cognitive science, thus AI. Gopnik [Gopnik1996], in a piece entitled "The Post-Piaget Era," expressed an apparently common view that Piaget's theories are increasingly implausible, and explored the various contenders for alternatives, to include the information processing approach. Yet in the same year, Lourenco and Machado [Lourenco1996] published a strong defense of Piagetian theory, arguing that its critics had approached the theory "from without, rather than from within," or to be more straightforward, that there was insufficient grasp of the theory and its implications, and that elementary mistakes in interpretation were being made.

Another conception may well be that connectionism has already accounted for Piaget's observations. Rogers and McClelland [Rogers2004], in support of their connectionist vision of the development of cognition, note that several connectionist models have been developed which can perform developmental learning tasks. One such task is the balance scale problem of Piaget. Here we have a small apparatus, like a small teeter totter, upon which the child can hang weights on each side to balance the "teeter totter" arm. The children gradually (in stages, over years, from roughly age 3 to age 7) learn the principle of torque, that is, the rule for balancing two differing weights (say a five-gram weight and a ten-gram weight) on each side, allowing compensation by varying the distances from the center of the scale. For example, a five-gram weight on the right, and two feet from the center, will balance a ten-gram weight on the left, only one foot from the center (length × weight = length × weight, or 5 lbs.' × 2 ft = 10 lbs.' × 1 ft). Yet

Quinlan et al., in a detailed analysis, have shown that in balance scale learning simulations, connectionist networks never in fact learn the principle of torque, nor do their internal weight representations ever approximate intermediate rules that correspond to the human phases of learning [Quinlan2007]. We will see, shortly, on looking more deeply into some Piagetian tasks, why this connectionist approach could only have been a very weak approximation to the human developmental trajectory.

Piaget seems, in large proportion, phenomenology. There is not a great deal of mechanism underlying his theory, mainly his insistence on the importance of action (assimilation, accommodation), and the growth and ability of intercoordinated actions to support mathematical group relations. It is a search for a "law of evolution," and as Piaget notes, "It is only in relation to evolution that such a law or attempt at differential analysis of behavior patterns acquires some meaning" ([Piaget1954], p. 381). This apparent reliance on phenomenology would generate criticism [Brainerd1978] that the theory is merely a *redescription* of manifest behavior with no explanatory value. Yet all this is being superseded by the awareness that in his "stages," Piaget is describing the necessary result of the evolution of a system characterized by nonlinear dynamics with its natural bifurcations (cf. [vandeer1992], [Raijmakers1996], [Molenaar2000]). We are looking at the natural result of a self-organizing system.

A COST TRAJECTORY—CAUSALITY, OBJECT, SPACE, AND TIME

To Piaget, this dynamic development in the brain's organization is underpinning the birth of a correlated set of fundamental concepts—causality, object, space, and time. Elsewhere [Robbins2009] this has been termed the "COST" of explicit memory, for simultaneously this complex is giving birth to, and required for, the explicit memory of events—in Piaget's terms, the ability to "localize events in time." This too underpins the development of the ability to symbolize.

As this is a literature to which, it could be suspected, AI does not pay much attention, but where the implications are profound, we are going to take a brief review of Piaget's description of the development of this ability through his six "stages." This story of the birth of the explicit is buried, implicitly so to speak, in his work, *The Construction of Reality in the Child* [Piaget1954]. While we've discussed that some have considered Piaget perfectly explainable via the computer-as-mind or information processing framework, we can be assured this is not the case. For its coherence, Piaget's vision of mind requires the Bergson-Gibson framework, its direct perception, and its temporal metaphysic.

The First Two Stages (0–4 months)

In the first two stages identified by Piaget, everything, he argues, takes place as though time were completely reduced to impressions of expectation, desire, success, or failure. The world virtually emanates from one's actions. "Objects" do not yet exist. There is the beginning of sequence linked with the development of different phases of the same act. But each sequence is a whole isolated from the others. Nothing yet enables the subject to reconstruct his own history and to consider these acts succeeding one another. Each sequence consists in a gliding from the preliminary phase of desire or effort, experienced as a present without a past. Finally, he argues, "this completely psychological *duration* is not accompanied by a seriation of events external and independent of the self—since *no boundary yet exists*" ([Piaget1954], p. 393, emphasis added). He comments, then, that the only form of memory evidenced in the first two stages is "… the *memory of recognition* in contradistinction to *the memory of localization or evocation*" (p. 369). He argues further that this recognition fails to transcend a basic sensation of the familiar, and this does not entail "… any clear differentiation between past and present but only the qualitative extension of the past into the present" (p. 369).

This latter comment on recognition—the "qualitative extension"—one would presume, derives from Piaget's mentor of sorts, Bergson, who saw one form of recognition as the familiarity produced by an "automatic motor accompaniment," say, as when we walk up our own driveway for the nth time (discussed further in Chapter 7, "Generative AI and Human Speech"). The "duration" in which the child dwells also is an invocation of Bergson's "duration"—the qualitative, nondifferentiable flow of time which cannot be defined as an abstract series of "instants," that is, the child (and the brain) does not dwell in the abstract "time" of the classic metaphysic.

Stage 3 (5–7 months)

Piaget describes in the next stages how actions lead to *localization in time*. Thus, he describes Laurent (at 8 months), seeing his mother enter the room and watching her until she seats herself behind him. Laurent resumes playing but turns around several times in succession to look at her again, even though there is no sound or noise to remind him she is there. This, Piaget notes, "…is the beginning of *object formation* analogous to what we have cited in connection with the third stage of objectification. This process is on a par with *a beginning of memory or localization in time*" (p. 375, emphasis added).

Piaget sees here an elementary concept of before and after. But Laurent's turning around is not yet "evocation," for it is by virtue of the movement of turning around in order to see that the child forms his nascent memory. But however "motor" and however little representative, it is the beginning of localization.

But is there an orderly arrangement of memories relating to external events? The child yet fails to note anything regarding the sequential positions of objects—the object has no spatial permanence. So it is that Laurent, immediately after the behavior pattern where he looks back repeatedly for his mother, reveals an action which clarifies this turning behavior's meaning, for when his mother leaves the room, "Laurent watches her until she reaches the door, then, as soon as she disappears, again looks for her behind him in the place where she was at first!" (p. 377).

Laurent's mother is not yet a permanent object, moving from place to place, but rather a memory image capable of reappearing precisely where it was previously perceived. The order is a function only of Laurent's action, as though, using Piaget's example, having laid my watch on the desk, then covered it with a manuscript, and now having forgotten these displacements of my watch, I look for it again in my pocket where I always reach for it. In essence, the child perceives the order of phenomena/events only when he himself has been the cause. *A universe without causality externalized in things cannot comprise a temporal series other than those relating to acts of the subject.*

Stage 4 (8–11 months)

In this stage, the before–after relation begins to be applied to the object, not just actions. The object disappears behind a screen, but while perceiving the screen, the child retains the image of the object and acts accordingly. The child is now recalling events, not merely actions. But this is extremely unstable. The object is hidden originally at A, and found by the child at A (e.g., under a pillow). The object is then moved in full view to B, and hidden, under another pillow (B). The child goes to B, but if the object is not found immediately, goes back to A, where the action was originally successful. The child's memory is now such that he can reconstruct short (but only short) sequences of events independent of the self.

The searching phenomenon described previously is commonly termed the A-not-B error in the developmental literature and has been the subject of extensive research (see reviews for example by [Markovitch1999] [Munakata1998]), including the development of dynamical mechanisms to explain it (Thelen et al., [Thelen2001]). In our opinion, the dynamic systems models fall well short, partially for the reason that A-not-B cannot be isolated from the whole developmental complex and trajectory—here being (only minimally) described—of which it is a part.[6]

[6.] Thelen et al. [Thelen2001] explained the A-not-B error by modeling the decision as a dynamic field which evolves continuously under the influence of the specifications of the task environment, the specific cue to reach A or B (which must be remembered), and after the first reach, the memory dynamics which bias the field (almost as though biasing the release of a mass-spring) on a subsequent reach. But when looking at the reflections of the authors in trying to account for the reasons for change of, and the origins of, the critical parameter value in their model (h, or cooperativity) that determines a reach to A or to B, one sees this: "They learn to shape their hands in anticipation of objects to be reached and then to differentiate the fingers to pick up small items ... They start to incorporate manual actions with locomotion such as crawling and walking. And they begin to have highly differentiated manual activities with objects of different properties, such as squeezing soft toys and banging noisy ones. It is a time of *active exploration of the properties of objects by acting on them* and of active exploration of space by moving through it" (p. 31, emphasis added). But this is exactly what we viewed Jacqueline, Lucienne and Laurent doing. The previous description simply strips out the embedding in the larger dynamic trajectory Piaget is trying systematically to describe. It is, yes, but a weak "redescription" (and attempt at *reduction*) of Piaget. The simple activities Thelen et al. note, when viewed as we have before from a larger perspective, are part of the emergence of causality, of sequence in time, of the self as object among other objects, of the body as a force among other forces, that is, an emerging complex of base concepts.

Stage 5 (12–18 months)

In this stage, Piaget notes that time definitely transcends the duration inherent in personal activity, which to say that it transcends our basic experience of continuous flow; it is approaching the conceptual, and it is "... applied to *things* themselves and to form the continuous systematic link which unites the events of the external world to one another" (p. 385, emphasis added).

There is now a systematic search for the vanished object, taking into account multiple displacements. An object hidden in *A*, found in *A*, hidden now in *B*, is no longer looked for in *A* (the former source of practical, action-issued success), but directly in *B* (although only for visible displacements). For the first time, the child seems capable of elaborating an "objective series."

Causality too is becoming "objectified." Such causality, on permanent objects, in an ordered space, entails the order of events in time. Jacqueline is watching a toy comprised of a revolving ball with chickens. The slightest movement of the ball puts the chickens in a pecking motion, something which Jacqueline notices when she touches the ball gently. She then systematically moves the ball as she watches the chickens, convinced now of the existence of a relationship not totally understood, and finally, each time the swinging stops completely, pushing the ball very delicately with her finger. "She definitely conceives the activity of the ball as causing that of the chickens ... Moreover, the ball is not, to her, a mere extension of her manual action" ([Piaget1954], p. 310).

The self has now become an object among other objects, just one source of force among other forces. Jacqueline, for example, places a red ball on the floor and waits for it to roll. Of course, the ball does not move, and after five or six attempts, she pushes it slightly. To Piaget, the ball is an autonomous center of forces, "... causality thus being detached from the action of pushing to be transferred onto the object itself" ([Piaget1954], p. 309).

Objects are seen in causal relation to other objects. Jacqueline wishes to have a plush cat on the floor, but not knowing how to

---

It will be more apparent shortly: Dynamic systems theory does not have the resources to adequately account for what is going on in Piaget's trajectories.

pull it, she simply touches it with her stick. Thus, to her, the spatial contact between the plush cat and the stick seems to her sufficient to displace the object. So, for her, "causality is spatialized" ([Piaget1954], p. 321) but the physics and mechanics needed have not yet been experienced or understood. Finally (two months later) she uses the stick correctly.

The sticks, the strings used to make objects move, and so on, are no longer only symbols of personal activity, but objects inserted into the web of events, therefore into conditions of time and place. Laurent, trying to reach an object, revolves a box serving as its support. The concepts "before/after" are no longer limited to her acts but are applied to the phenomena—to their displacements, perceived and remembered.

Stage 6 (19+ months)

Finally, *the symbolic is itself emerging*, where the symbol is at the level of action. Piaget is playing with his 16-month-old daughter Lucienne and has hidden an attractive watch chain inside a matchbox, reducing the opening to a very small slit. Lucienne tries to open it with two "schemas" or plans for action she already possesses, turning the box over to empty it and then attempting to slide her finger into the slit and extract the coveted item. Both fail. "She looks at the slit with great attention; then, several times in succession she opens and shuts her mouth, at first slightly, and then wider and wider ..." ([Piaget1952], p. 338). Not yet able to think clearly with images, the mouth opening—this simple motor action or motor image—is a "signifier" or symbol. Quickly thereafter, Lucienne unhesitatingly puts her finger into the slit and pulls to enlarge the opening.

The child can now handle even invisible displacements, and Piaget notes that the "elaboration of the temporal field" *requires the development of images*, that is, representations. This lack of representative imagery is why the earlier remembered series are so short. These operations, he notes, are no longer purely practical memory of a primitive series (as a mother coming into the room, an object moved under a pillow), but rather, are nothing other than *evocative memory*, "... making it possible to *locate in time the actions of the self-amidst the other events*" ([Piaget1954], p. 392, emphasis added).

THE BIRTH OF THE SYMBOLIC

If we sum up what is being said about the development of explicit memory, it might be said simply: there is no full explicit memory without at least some COST. Here COST stands for causality, object, space, and time. Together these concepts form an interrelated, supporting group, and together they grow from intercoordination of actions. As Piaget describes it from the perspective of the development of one of these, namely from the concept of the object, from a mere extension of the child's activity, the object is gradually dissociated from activity. Resistance initiates this dissociation, for example, obstacles or complications of the field of action as in the appearance of a screen obscuring the favorite toy ([Piaget1954], p. 103). Action gradually becomes a factor among other factors, and the child comes to treat their own movements on a par with those of other bodies. When "things" are detached from actions, and actions placed within the network of surrounding events (or "series" in Piaget's terms), the subject is enabled to construct a system of relations to understand himself in relation to them. To organize such series "... is to form simultaneously a spatiotemporal network and a system consisting of substances and of relations of cause and effect ... Hence the construction of the object is inseparable from that of space, time, and causality" ([Piaget1954], p. 103).

The emergence of COST, Piaget argues, is the event that supports true explicit memory. This occurs gradually, emerging in Piaget's sixth stage, at 18–24 months. Simultaneously, then, this dynamical trajectory is driving toward liberation from perception, going beyond the present as Piaget insists ([Piaget1954], p. 390), toward *representation by an image*, emerging only at the sixth stage. So, Piaget explains, "That is why the temporal series just described are revealed as so short and so dependent on the constructions characteristic of object, space, and causality; *it is why, for lack of representations properly so called ...*" ([Piaget1954], p. 391, emphasis added).

The "objective series," where the permanence and displacements of objects are gradually embodied in practical action, are now extended as the *representative* series. It is the same operations, now performed on the mental, the representative level. To Piaget, this is nothing other than evocative memory, but equally, this evocative memory is not a special faculty, but only mental or "reproductive assimilation" (i.e., a relation of object–images to actions), to the extent that it constructs mentally, not in the physical world, a more extensive past. And indeed, this reproductive ability had to be present from the beginning; it is the fundamental basis of explicit memory; it is the ability of the brain to redintegrate a past experience. *There is no necessary implication here at all that the events of an infant's or young child's past are lost, or never stored.*

The resultant of this dynamical development toward COST that Piaget is trying to describe is a *representative* form, the ability to treat these events as *symbols* within a localized past. Thus, we see a telling incident where Jacqueline (19 months) picks up a blade of grass that she puts in a pail as if it were one of the grasshoppers a little cousin brought her a few days before, saying, "Totelle [sauterelle, or grasshopper] totelle, jump, boy [her cousin]." This is now *explicit* memory. As Piaget states, "In other words, perception of an object which *reminds her symbolically of a grasshopper* enables her to evoke past events and reconstruct them in sequence" ([Piaget1954], p. 391, emphasis added).

The developed explicit memory, the symbolic function and COST are all interrelated, mutually supportive. The infant brain can create neural patterns redintegrating past events until the cows come home, but until these past events can become *symbolic*, until they take their place within a conceptual past, they are little more than present phantoms. So, for Jacqueline, the present event (the piece of grass) now functioned within what we term an *articulated simultaneity*, for it served both as itself and as a symbol of a past event, assuming both one meaning then another, yet within a global, temporal whole or state of consciousness, all underpinned by the indivisible flow of time in which the brain dwells. Explicit memory is bound to the development of the symbolic—the ability to employ a form (the piece of grass) abstracted from our experience to represent another aspect of experience in the past (the grasshopper) what Cassirer would term the "symbolic function" [Cassirer1929]. The symbolic function carries this aspect of an articulated simultaneity— this "oscillation" between two elements within a temporal whole. This tells us that the symbolic itself is a problem in the relation of mind to time.

All of the preceding, if one looks closely, comprise the framework in which the famous "symbol grounding" problem is being resolved, albeit a framework that requires Bergson's "device." The blade of grass, for Jacqueline, is surely a *grounded* symbol, grounded that is for a 4-D being whose entire concrete experience is retained, is accessible, and wherein that experience can now be accessed within the resultant organizational framework provided by the trajectory toward COST (a trajectory, we have failed to note thus far, which clearly requires development of some form of coordinative dynamics with the prefrontal cortex in concert with the hippocampus). To see the full, dynamic complexity of the picture, however, there is still required, the simultaneity feature noted previously.

THE SIMULTANEITY OF THE SYMBOLIC STATE—CASSIRER

We need to underline the critical aspect of this symbolic representation, namely the articulated simultaneity. It is not only the events of the past that become symbolic. For Cassirer it is the events of perception as well. The pathologies, he thought, indicate that, "… the contents of certain sensory spheres seem somehow to lose their power of functioning as pure means of representation …" ([Cassirer1929], p. 236). Some aphasics cannot make a simple sketch of their room, marking in it the positions of objects. Many patients can orient themselves on a sketch if the basic schema is already laid down, for example, the doctor prepares the sketch and indicates by an X where the patient is. But the truly difficult operation is the spontaneous choice of a plane as well as a center of coordinates. Thus, one of Head's [Head1926] patients would express his problem as the "starting point, but once it was given him everything was much easier."

The same principle operates in the aphasic's dealing with number and time. One patient could recite the days of the week or the months of the year, but given an arbitrary day or month, could not state what came before or after. Although he could recite the numbers in order, he could not count a quantity. Given a set of things to count, he could not progress in order, but frequently went back again. If he had arrived, for example, at "six," he had no comprehension that he had a designation for the quantity thus far achieved, that is, a cardinal number. When asked which of two numbers are larger, for example, 12 or 25, many aphasics can do so only by counting through the whole series, determining that in this process the word 25 came after the word 12.

As Cassirer notes, "Where quantity no longer stands before us as a sharply articulated multiplicity, it cannot be strictly apprehended as a unity, as a whole built up of parts" ([Cassirer1929], p. 250). But to achieve this (and we shall see this again shortly in the seriated set of flows of Piaget), every number must carry a *dual role*. Thus, to find the sum of 7 + 5, or the difference 7 - 5, the decisive factor is that the number 7, while retaining its position in the first series of 7, now is taken in a new meaning, becoming the starting point of a new series where it assumes the role of zero (Figure 4.4). Again, as in the representation of space, we have a free positing of a center of coordinates. "The fundamental unities must be kept fixated, but precisely in this fixation be kept mobile, so *that it remains possible to change from one to the other*" ([Cassirer1929], p. 250, emphasis added). The number 7 must maintain its meaning as 7, yet simultaneously assume the meaning of zero. This is a pure problem of representation in time. More precisely, as we will discuss subsequently, it is a problem of representation in an extended time supporting true simultaneity.

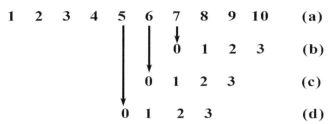

FIGURE 4.4. The problems: 7 + 3 = 10, 6 + 3 = 9, 5 + 3 = 8. The number 7 functions simultaneously as zero in (b), 6 functions as zero in (c), and 5 as zero in (d). (After Cassirer, 1929/1957).

This extends to the sphere of action, in the apraxias. One patient (Gelb and Goldstein [Gelb1918]) could knock on the door if the door was within reach, but the movement, although begun, was halted at once if he was asked to move one step away from the door. He could hammer a nail if, hammer in hand, he stood near the wall, but if the nail was taken away and he was asked to merely indicate the act, he was frozen. He could blow away a scrap of paper on the table, but if there was no paper, he could not blow. This is not the loss of memory images, as was once widely held. He just blew the paper; how could the image have been lost? It is not a failure to create a sensuous "optical space" as Gelb and Goldstein thought. He is staring at the door, and still cannot act. It is the inability to create an abstract, free space for these movements.

Thus, Cassirer notes, "For this latter is the product of the "productive imagination": it demands an ability *to interchange present and nonpresent*, the real and the possible" ([Cassirer1929], p. 271, emphasis added). As normal individuals, we can perform the hammering a nail just as well into a merely imagined wall as a real wall, because in free activity we can vary the elements sensuously given, and by thought we can "… exchange the here and now with something else that is not present … [this] requires a schematic space" ([Cassirer1929] p. 271).

This extends to analogy. Thus, the psychic blindness patient of Gelb and Goldstein was utterly unable to comprehend linguistic analogy or metaphor. He made no use of either in his speech, in fact, rejected them entirely. He dealt only in realities. Asked by Cassirer, on a bright, sunny day, to repeat, "It is a bad, rainy weather day," he was unable to do so. Again, analogy requires precisely the same achievement of representation. To say, "The spoon is a catapult" (after using it to launch a pea), the spoon must be taken simultaneously in two different modes. One must place oneself simultaneously now in one, now in another meaning, yet maintain a vision of the whole. It is this "articulation of one and the same element of experience with different, equally possible relations, and simultaneous orientation in and by these relations, [that] is a basic operation essential to thinking in analogies as well as intelligent operation with numbers …" ([Cassirer1929], p. 257).

It is this simultaneous articulation that supports little Jacqueline and her "grasshopper." Again, as Cassirer argued, there is an interchange of present and non-present, that is, of present and past. The little piece of grass can now be a symbol; it can be simultaneously both a piece of grass—a perceptual image—and a "grasshopper" which once she saw hopping around. But now we find ourselves inevitably relating this to the problem of Weiskrantz and his "product" (present × past event). When H.M. looks at the mirror tracing apparatus, should it not be simultaneously both the present apparatus and the past apparatus upon which he once worked? Sitting on the porch, the wind chimes instantly become a "symbol," simultaneously both of themselves and the memory of buying other chimes for one's wife for an anniversary. All such surrounding objects can become symbolic of the past. But now we see this apparently simple ability through the underlying complexity of the dynamic that must support it.

THE DYNAMICAL LENS AND SIMULTANEITY IN PIAGET

It is worth another look at Piaget's developmental trajectory, further along, where he is implicitly relying on the dynamical simultaneity Cassirer is pointing to. This point also illustrates what can be termed the *dynamical lens on past events* inherent in Piaget's view, that is, as noted in the symbol grounding comment, wherein that experience can now be accessed within the resultant organizational framework provided by the trajectory toward COST. Piaget describes a simple memory experiment with children aged 3 to 8 [Piaget1968]. They are shown a configuration of ten small sticks (Figure 4.5 [A]). They are asked to have a good look so they can draw it again later. One week later, without having seen the configuration again, they are asked to draw what they were shown before. Six months later they are asked to do the same thing.

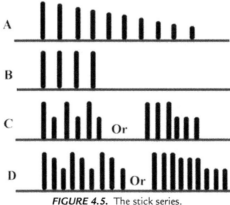

FIGURE 4.5. The stick series.

In the one-week interval case, the reconstruction of the event is dependent, in Piaget's terms, on the "operational schemata" to which the child assimilated the event. Memory is dependent, in other words, on the child's current ability to coordinate actions. At approximately age 3–4, the series is reproduced as in Figure 4.5[B]. Slightly older children (4–5) remember the form in Figure 4.5[C]. Figure 4.5[D] is a slightly more advanced reproduction, while at 6–7, the child remembers the original series. After six months, as Piaget describes, children of each age group claimed they remembered very well what they had seen, yet the drawing was changing. The changes generally, with rare exceptions, moved in small jumps, for example, from A to C, or from C to D, or from D to A.

The drawing of a series—the process of seriation—requires what Piaget terms "concrete operations" (to be discussed in Chapter 5, *Conscious* Cognition). To order a simple series, A, B, C (where A is longer than B, and B is longer than C), one must *simultaneously relate or coordinate the height of B relative to A*, the height of B relative to C. This is fundamentally based on the intercoordination of action. As the child follows this developmental trajectory, Piaget argues, the available dynamics support successively different "decodings" of the memory, that is, the events are reconstructed with an increasingly sophisticated logical structure. The past event is seen through this "dynamical lens." But simultaneously then, when the series is a perceived event, it is being perceived through this same dynamical lens. The events gain an increasingly complex symbolic structure.

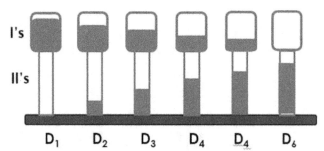

FIGURE 4.6. Two flows at different velocities. In a series of stages, the water is gradually emptied from the top beaker (I), while the lower beaker (II) gradually fills.

This is but one of a large number of such examples of events treated by Piaget where we see this developing dynamical lens, and the need for a simultaneous relation of magnitudes. In another, more dynamic event (*The Child's Conception of Time* [Piaget1927]), the experimental apparatus consists simply of two differently shaped flasks, one placed atop the other, with a tap or valve between the two (Figure 4.6). The top flask is initially filled with liquid. The child (ages 5–9) is provided with a set of drawings of the two flasks (with no liquid levels filled in). At regular intervals or stages a fixed quantity of liquid is allowed to run from the top flask into bottom until the top is empty. Because the flasks are differently shaped, the one empties at a different velocity than that at which the other fills. At each stage, starting at the beginning, the child is asked, using a fresh drawing, to draw a line on each flask indicating the level of the liquid.

The drawings are now shuffled, and questions ensue:

1. The child is asked to reconstruct the series, putting the first drawing made down, the next to the right, and so on.

2. Every sheet is now cut in half and the drawings shuffled. The child is again asked to put them in order.

The children go through stages again. At first, they cannot order the uncut drawings (the Ds of Figure 4.6). Then they achieve this (uncut) ordering but fail ordering the cut drawings. The child may come up with an arrangement of the cut drawings such as I_3, I_1, I_2, I_5, I_6 above II_1, II_5, II_6, II_3, II_2. Even with coaching, he cannot achieve a correct order. Beneath these failures lies a certain mental rigidity. The child is required here to move mentally against the irreversible, experiential flow of time. Regardless of the irreversible flow of the water, they must perform a reversible operation. They must construct a series $A \Rightarrow B \Rightarrow C$ ... where the arrows carry a dual or simultaneous meaning standing for "precedes" as well as "follows." It is a pure problem of representation. The child must also grasp the causal connections within and between the two flows. Ultimately, the children will surely and securely order this double series with its causal link as the operating principle. The operations involved in this coordination of two motions, participating in the logic of objects we call causality, underlie our "schema" of time. This schema, supporting our memory operations, our retrieval of events, is again a lens upon our experience. It is a cognitive capability that develops over several years, based on the coordination of actions, first sensorimotor, then mental, and something, we might note, that must be subject to disruption via a traumatic injury or disease.

NOTE ON THE IMAGE AND THE "IMAGE MEMORY"

Bergson has been criticized as though his memory consists of a set of images. Sartre [Sartre1962], who totally failed to understand Bergson, the key, holographic aspect totally passing Sartre by, was such a critic. Even Piaget, a student in at least a class or so of Bergson, obviously influenced but concerned to distance himself [Piaget1965], made the same critique. But there is the virtual, but there is no static memory of images. Virtual experience must pass through the dynamic lens provided by the brain, only as such coalescing as an image or imagery. Bergson made this very clear in *Matter and Memory*: "Pure memory, though independent in theory, manifests itself as a rule only in the colored and living image which reveals it," and once this is realized, we see that Piaget is describing the development and nature of this dynamic lens.

The physical organism spends several years and a great deal of effort following this "law of evolution" or developmental trajectory to produce this dynamical possibility. In the sphere of explicit memory, and in the spheres, Cassirer discussed—sketching, analogy, numbers, time, voluntary actions—an essential feature of the underlying dynamic in each is the articulated simultaneity supporting the symbolic nature of these functions. If it took a dynamic, neural developmental trajectory of two years to support this ability, it would seem logical this same neural organization can be undone—damaged—and amnesia is one such effect, and more on this has been said elsewhere [Robbins2009]. But the import of this trajectory for AI is the critical issue here. If it requires two years of dynamic, organizational development to achieve explicit memory, which is simultaneously to say to achieve the symbolic, how, one can ask, can it ever be said that a robot, a robot driven by a GPT, a robot/GPT that has engaged in nothing like this trajectory even in a supposedly sped up time frame, is in any way equivalent to the human mind? How could Connor Leahy's statement, "I think that human beings are approximating what GPT-3 is doing and not vice versa" [Thompson2023, at 5:15], be given any credence when the implications of this two-year trajectory have not even been considered?

We shall see (Chapter 5) that imagery is essential in Piaget's theory of cognitive development, the heart of his "concrete operations." In the ability to represent, to support symbolic relations, to employ imagery derived from past experience even in this simple operation, for example, Jacqueline's grasshopper, we are back to something noted in Chapter 1, namely, that AI, and cognitive science as well, have no place in their framework for imagery. For Pylyshyn, it was peskily there (but *how*, given he cannot explain the perceptual image in the first place) but the image was just redundant, useless; data structures and logical operations handle all of cognition. Add in too that consciousness is not seen as necessary as well. Now we look even further into why this is all misconceived, that is, why both imagery and consciousness are essential to cognition.

REFERENCES

[Anderson1977] Anderson, J. A., Silverstein, J. W., Ritz, S. A., & Jones, R. S. "Distinctive features, categorical perception, and probability learning: Some applications of a neural model," *Psychological Review* (1977): *84*(5), 413–451.

[Bergson1902] Bergson, H. "Intellectual effort," in *Mind-Energy, Lectures and Essays*. (W. Carr, Trans.) New York: Henry Holt and Co, 1920. (Originally published in the *Revue Philosophique*, January 1902.)

[Bingham1993] Bingham, G. P. "Perceiving the size of trees: Form as information about scale," *Journal of Experimental Psychology: Human Perception and Performance* (1993): *19*(6), 1139–1161.

[Block1995] Block, N. "On a confusion about the function of consciousness," *Behavioral and Brain Sciences* (1995): *18*(2), 227–247.

[Brainerd1978] Brainerd, C. J. "The stage acquisition in cognitive developmental theory," *Behavioral Brain Sciences* (1978): *1*(2), 173–213.

[Buzsáki2017] Buzsáki, G. *The Brain from Inside Out*. Oxford: Oxford University Press, 2017.

[Cassanto2015] Casasanto, D., & Lupyan, G. "All concepts are ad hoc concepts," in E. Margoulis & S. Lawrence (Eds.), *The Conceptual*

Mind: New Directions in the Study of Concepts (pp. 543–566). Cambridge, MA: MIT Press, 2015.

[Cassirer1929] Cassirer, E. *The Philosophy of Symbolic Forms, Vol. 3: The Phenomenology of Knowledge.* New Haven, CT: Yale University Press, 1929/1957.

[Chalmers1992] Chalmers, D. J., French, R. M., & Hofstadter, D. "High level perception, representation and analogy: A critique of artificial intelligence methodology," *Journal of Experimental and Theoretical Artificial Intelligence* (1992): 4(3), 185–211.

[Craig2000] Craig, C. M., & Bootsma, R. J. "Judging time to passage," in M. A. Grealy & J. A. Thomson (Eds.), *Studies in Perception and Action V.* Mahwah, NJ: Erlbaum, 2000.

[Crowder1993] Crowder, R. G. "Systems and principles in memory theory: Another critique of pure memory," in A Collins, S. Gathercole, M. Conway, & P. Morris (Eds.), *Theories of Memory.* Mahwah, NJ: Erlbaum, 1993.

[Dietrich2000] Dietrich, E. "Analogy and conceptual change, or you can't step into the same mind twice," in E. Dietrich & A. B. Markman (Eds.), *Cognitive Dynamics: Conceptual and Representational Change in Humans and Machines* (pp. 265–294). Mahwah, NJ: Erlbaum, 2000.

[Domini2002] Domini, F., Vuong, Q. C., & Caudek, C. "Temporal integration in structure from motion," *Journal of Experimental Psychology: Human Perception and Performance* (2002): 28(4), 816–838.

[Doumas2008] Doumas, L., Hummel, J., & Sandhofer, C. "A theory of the discovery and predication of relational concepts," *Psychological Review* (2008): 115(1), 1–43.

[Eich1985] Eich, J. "Levels of processing, encoding specificity, elaboration, and CHARM,"

Psychological Review (1985): 92(1), 1–38.

[French1990] French, R. M. "Sub-cognition and the limits of the Turing Test." *Mind* (1990): 99(393), 53–65.

[French1999] French, R. M. "When coffee cups are like old elephants, or why representation modules don't make sense," in A. Riegler, M. Peshl, & A. von Stein (Eds.), *Understanding Representation in the Cognitive Sciences* (pp. 158–163). New York: Plenum, 1999.

[Freyd1984] Freyd, J. J., & Finke, R. A. "Representational momentum," *Journal of Experimental Psychology: Learning, Memory and Cognition* (1984): *10*(1), 126–132.

[Freyd1987] Freyd, J. J., "Dynamic mental representations," *Psychological Review* (1987): *94*(4), 427–438.

[Freyd1990] Freyd, J. J., Kelly, M. H., & DeKay, M. L. "Representational momentum in memory for pitch," *Journal of Experimental Psychology: Learning, Memory and Cognition* (1990): *16*(6), 1107–1117.

[Fuster1994] Fuster, J. M. "In search of the engrammer. Response to Eichenbaum et al.," *Behavioral and Brain Sciences* (1994): *17* (3), 476.

[Gayler2003] Gayler, R. W., "Vector symbolic architectures answer Jackendorf's challenges for cognitive neuroscience," in Peter Slezak (Ed.), ICCS/ASCS International Conference on Cognitive Science (pp. 133–138). Sydney, Australia: University of New South Wales, 2003.

[Gelb1918] Gelb, K., & Goldstein, K. "Zur psychologie des optischen wahrnehmungs- und erkennungsvorgangs," *Zeitschrift fur die gesamte Neurologie und Psychiatrie*, 41, 1918.

[Gelernter1994] Gelernter, D. *The Muse in the Machine: Computerizing the Poetry of Human Thought*. New York: Free Press, 1994.

[Gentner1983] Gentner, D., "Structure-mapping: A theoretical framework for analogy," *Cognitive Science* (1983): *7*(2), 155–170.

[Gick1983] Gick, M. & Holyhoak, K. "Schema induction and analogical transfer," *Cognitive Psychology* (1983): *15*, 1–38.

[Goldinger1998] Goldinger, S. "Echoes of echoes? An episodic theory of lexical access," *Psychological Review* (1998): *105*(2), 251–279.

[Gopnik1996] Gopnik, A., "The post-Piaget era," *Psychological Science* (1996): *7*(4), 221–225.

[Gregory1987] Gregory, R. *The Oxford Companion to Mind*. Oxford: Oxford University Press, 1987.

[Gunther2012] Gunter, K. *Mind, Memory, Time*. Carl Gunther, 2012.

[Guzik2023] Guzik, E. "OpenAI's GPT4 outperforms human originality in renowned creativity test," *CISION PR Newswire*, May 18, 2023. https://www.prnewswire.com/news/dr.-erik-guzik/

[Head1926] Head, H. *Aphasia and Kindred Disorders of Speech (Vols. I and II)*. Cambridge: Cambridge University Press, 1926.

[Hintzman1986] Hintzman, D. L. "Schema abstraction in a multiple-trace memory model," *Psychological Review* (1986): 93(4), 411–428.

[Hofstadter2013] Hofstadter, D. & Sander, E. *Surfaces and Essences: Analogy as the Fuel and Fire of Thinking.* New York: Basic Books, 2013.

[Howe1993] Howe, M., & Courage, M. "The emergence and early development of autobiographical memory," *Psychological Review* (1993): 104(3), 499–523.

[Indurkyha1999] Indurkyha, B. "Creativity of metaphor in perceptual symbol systems," *Behavioral and Brain Sciences* (1999): 22(4), 621–622.

[Jenkins1978] Jenkins, J. J., Wald, J., & Pittenger, J. B. "Apprehending pictorial events: An instance of psychological cohesion," *Minnesota Studies of the Philosophy of Science* (1978): 9, 129–163. https://conservancy.umn.edu/bitstream/handle/11299/185329/9_07Jenkinsetal.pdf

[Klein1970] Klein, D. B. *A History of Scientific Psychology.* New York: Basic Books, 1970.

[Kingma2004] Kingma, I., van de Langenberg, R., & Beek, P. "Which mechanical invariants are associated with the perception of length and heaviness on a nonvisible handheld rod? Testing the inertia tensor hypothesis," *Journal of Experimental Psychology: Human Perception and Performance* (2004): 30(2), 346–354.

[Kugler1987] Kugler, P. & Turvey, M. *Information, Natural Law, and the Self-Assembly of Rhythmic Movement.* Hillsdale, NJ: Erlbaum, 1987.

[Lashley1950] Lashley, K. "In search of the engram," in J. F. Danielli & R. Brown (Eds.), *Physiological Mechanisms in Animal Behaviour* (pp. 454–482). New York: Academic Press, 1950.

[Lourenco1996] Lourenco, O., & Machado, A. "In defense of Piaget's theory: A reply to 10 common criticisms," *Psychological Review* (1996): 103(1), 144–164.

[Markovitch1999] Markovitch, S. & Zelaso, P. "The A-not-B error: Results from a logistic meta-analysis," *Child Development* (1999): 70(6), 1297–1313.

[Mayes2001] Mayes, A. R., & Roberts, N. "Theories of episodic memory," in A. Baddeley, M. Conway, & J. Aggleton (Eds.), *Episodic Memory:*

New Directions in Research. New York: Oxford University Press, 2001.

[McClelland1995] McClelland, J. L., McNaughton, B. L., & O'Reilly, R.C. "Why there are complementary learning systems in the hippocampus and neocortex: Insights from the successes and failures of connectionist models of learning and memory," *Psychological Review* (1995): *103*(3), 419–457.

[McGeoch1942] McGeoch, J. A. *The Psychology of Human Learning*. New York: Longmans, Greene, 1942.

[McGuire2016] Maguire, E. A., Intraub, H., & Mullally, S. L. "Scenes, spaces, and memory traces: What does the hippocampus do?" *Neuroscientist* (2016); *22*(5), 432–439.

[Molenaar2000] Molenaar, P., & Raijakers, M. "A causal interpretation of Piaget's theory of cognitive development: Reflections on the relationship between epigenesis and non-linear dynamics," *New Ideas in Psychology* (2000): *18*(1), 41–55.

[Moscovitch2016] Moscovitch, M., Cabeza, R., Winocur, G., & Nadel, L. "Episodic memory and beyond: The hippocampus and neocortex in transformation," *Annual Review of Psychology* (2016): 67, 105–134.

[Munakata1998] Munakata, Y. "Infant perseveration and implications for object permanence theories: A PDP model of the A-not-B task," *Developmental Science* (1998): *1*(2), 161–184.

[Murdock1982] Murdock, B. B. "A theory for the storage and retrieval of item and associative information," *Psychological Review* (1982): *89*(6), 609–626.

[Paivio1971] Paivio, A. *Imagery and Verbal Processes*. New York: Holt, Rinehart, and Winston, 1971.

[Paivio1994] Paivio, A., Walsh, M., & Bons, T. "Concreteness effects on memory: When and why?" *Journal of Experimental Psychology: Learning, Memory and Cognition* (1994): *20*(5), 1196–1204.

[Pearsall1999] Pearsall, P. *The Heart's Code: Tapping the Wisdom and Power of Our Heart Energy*. New York: Harmony, 1999.

[Piaget1927] Piaget, J. *The Child's Conception of Time*. New York: Basic Books, 1927/1969.

[Piaget1952] Piaget, J. *The Origins of Intelligence in Children*. New York: International Universities Press, 1952.

[Piaget1954] Piaget, J. *The Construction of Reality in the Child*. New York: Ballentine, 1954.

[Piaget1965] Piaget, J. *Insights and Illusions in Philosophy*. New York: World Publishing, 1965/1971.

[Piaget1968] Piaget, J. *On the Development of Memory and Identity*. Clark University Press, 1968.

[Piantadosi2023] Piantadosi, S. "Modern language models refute Chomsky's approach to language," LingBuzz, 2023.https://lingbuzz.net/lingbuzz/007180

[Pittenger1975] Pittenger, J. B., & Shaw, R. E. "Aging faces as viscal elastic events: Implications for a theory of non-rigid shape perception," *Journal of Experimental Psychology: Human Perception and Performance* (1975): *1*(4): 374–382.

[Quinlan2007] Quinlan, P., van der Maas, H., Jansen, B., Booij, O., & Rendell, M."Re-thinking stages of cognitive development: An appraisal of connectionist models of the balance scale task," *Cognition* (2007): *103*(3), 413–459.

[Raijmakers1996] Raijmakers, M. E. J., van Der Maas, H. L. J., & Molenaar, P.C.M. "On the validity of simulating stagewise development by means of PDP networks: Application of catastrophe analysis and an experimental test of rule-like network performance," *Cognitive Science* (1996); *20*(1), 101–136.

[Rakison2007] Rakison, D. "Is consciousness in its infancy in infancy?" *Journal of Consciousness Studies* (2007): *14*(9–10), 66–89.

[Robbins1976] Robbins, S. E. "Time and Memory: The Basis for a Semantic-Directed Processor and its Meaning for Education." Doctoral Thesis, University of Minnesota, 1976.

[Robbins2006] Robbins, S. E. "On the possibility of direct memory," in V. W. Fallio (Ed.), *New Developments in Consciousness Research* (pp. 1-64). New York: Nova Science, 2006.

[Robbins2008] Robbins, S. E. "Semantic redintegration: Ecological invariance. Commentary on Rogers, T. & McClelland, J, Précis *Semantic Cognition: A Parallel Distributed Processing Approach*," *Behavioral and Brain Sciences* (2008): 726–727.

[Robbins2009] Robbins, S. E. "The COST of explicit memory," *Phenomenology and the Cognitive Sciences* (2009): *8*(1), 33–66.

[Robbins2017] Robbins, S. E. "Analogical reminding and the storage of experience: The Hofstadter-Sander Paradox," *Phenomenology and the Cognitive Sciences* (2017): 16, 355–385.

[Robbins2020] Robbins, S. E. "Is experience stored in the brain? A current model of memory and the temporal metaphysic of Bergson." *Axiomathes* (2020): *31*(2), 15–42.

[Rogers2004] Rogers, T., & McClelland, J. *Semantic Cognition: A Parallel Distributed Processing Approach*. Cambridge, MA: MIT Press, 2004.

[Rogers2008] Rogers, T., & McClelland, J. "Precis of Semantic Cognition, a parallel distributed processing approach," *Behavioral and Brain Sciences* (2008): *31*(6), 689–749.

[RoyalInstitution2014] The Royal Institution. "The Neuroscience of Memory—Eleanor Maguire," March 13, 2014, YouTube video. https://www.youtube.com/watch?v=gdzmNwTLakg

[Sacks1987] Sacks, O. *The Man Who Mistook His Wife for a Hat*. New York: Harper and Row, 1987.

[Sartre1962] Sartre, J. *Imagination: A Psychological Critique*. (F. Williams, Trans.). Ann Arbor, MI: University of Michigan Press, 1962.

[Semon1909] Semon, R. *Mnemic Psychology*. (B. Duffy, Trans.). Concord, MA: George Allen & Unwin, 1909/1923.

[Spaulding1996] Spaulding, T, & Murphy, G. "Effects of background knowledge on category construction," *Journal of Experimental Psychology: Learning, Memory, and Cognition* (1996): 22(2), 525-538.

[Sheldrake2012] Sheldrake, R. *Science Set Free. 10 New Paths to Discovery*. New York: Deepak Chopra, 2012.

[Sixty2018] "People who remember every second of their life: Total recall." *60 Minutes Australia,* September 21, 2018, YouTube video. https://www.youtube.com/watch?v=hpTCZ-hO6iI

[Teyler1986] Teyler, T. J., & DiScenna, P. "The hippocampal memory indexing theory," *Behavioral Neuroscience*, (1986): *100*(2), 147–152.

[Thelen2001] Thelen, E., Schoner, G., Scheler, C., & Smith, L. "The dynamics of embodiment: A field theory of infant perseverative reaching," *Behavioral and Brain Sciences* (2001): *24*(1), 1–86.

[Thompson2023] Thompson, A. D. "Countdown to AGI: 42% in March 2023 (Transformer, GPT-3, TPUv4, H100, ChatGPT embodied, PaLM-E)," March 11, 2023. https://www.youtube.com/watch?v=qPI8fB2XL3w[Turvey1995] Turvey, M., & Carello, C., "Dynamic Touch," in W. Epstein and S. Rogers (Eds.), *Perception of Space and Motion*, San Diego: Academic Press, 1995.

[UCICNLM2018] UCI CNLM. "How Thinking About Memory Has Changed in the Past 35 Years," July 18, 2018. YouTube video. https://www.youtube.com/watch?v=gYvhMTq2-Ac

[Vandeer1992] Van deer Maas, H. L .J, & Molenaar, P. C. M. "Stagewise cognitive development: An application of catastrophe theory," *Psychological Review* (1992): 99(3), 395–417.

[Verbrugge1977] Verbrugge, R. "Resemblances in language and perception," in R. E. Shaw & J. D. Bransford (Eds.), *Perceiving, Acting, and Knowing*. Hillsdale, NJ: Erlbaum, 1977.

[Vicente1998] Vicente, K. J., & Wang, J. H. "An ecological theory of expertise effects in memory recall," *Psychological Review* (1998): *105*(1), 33–57.

[Vicente2000] Vicente, K. J. "Revisiting the constraint attunement hypothesis: Reply to Ericsson, Patel and Kintsch, and Simon and Gobet, (2000): *Psychological Review, 107*(3), 601–608.

[Weiskrantz1997] Weiskrantz, L. *Consciousness Lost and Found*. New York: Oxford, 1997.

[Zimmer2000] Zimmer, H. D., Helstrup, T., & Engelkamp, J. "Pop-out into memory: A retrieval mechanism that is enhanced with the recall of subject-performed tasks," Journal *of Experimental Psychology: Learning, Memory and Cognition* (2000): *26*,3, 658–670.

[Zimmer2001a] Zimmer, H. D., & Cohen, R. L., "Remembering actions: A specific type of memory?," in H. D. Zimmer & R. L. Cohen (Eds.), *Memory for Actions: A Distinct Form of Episodic Memory?* (pp. 3–24). Oxford: Oxford University Press, 2001.

[Zimmer2001b] Zimmer, H. D., "Why do actions speak louder than words?: Action memory as a variant of encoding manipulations or the result of a specific memory system?," in H. D. Zimmer & R. L. Cohen (Eds.), *Memory for Actions: A Distinct Form of Episodic Memory?* (pp. 151–198). Oxford: Oxford University Press, 2001.

CHAPTER 5

CONSCIOUS COGNITION

THE ABSTRACT AND THE CONCRETE

Pylyshyn, as we saw in Chapter 1 ("Introduction: AGI—The Gleam in the Eye of AI") while ultimately allowing that mental images might exist, challenged the cognitive science field to show why they are needed. He said "In other words, the image has lost all its picture-like qualities and has become a data-structure ... In fact, it [the image] can be put directly into one-to-one correspondence with a finite set of propositions" ([Pylyshyn1973], p. 22).

At this point, where we see "seeing the image" reduced to things like "testing the identity of symbols," we can see that Pylyshyn, and subsequently cognitive science, were quite content to have no theory of the image of the external world, that is, of our *experience*, and not a prayer of handling Chalmers' future and only partially stated version of the problem in terms of qualia. It was acceptable, although hardly admitted, that the computer would be blind, however happily testing symbols. But this assertion, "... it [the image] can be put directly into one-to-one correspondence with a finite set of propositions," is what we want to start with here.

Back we go to stirring coffee with the spoon. This is a *dynamic image*—our ever-changing *experience*—and it is filled, structured, with invariance laws. "The surface is swirling." This is surely one of those "propositions" in Pylyshyn's "finite set" in one-to-one correspondence with this dynamic event. How is this proposition (or a set of such) in one-to-one correspondence with the swirling which is dynamic—a constant motion, a dynamic form, a radial flow field? Further, these are *multimodal* invariants; they are coordinate with each other over this flow. The swirling itself is in sync with, in correspondence with, the periodic motion of the spoon—an adiabatic invariant, defined over a kinesthetic flow field involving the hand-arm and coordinate with the "clinking" of the spoon as it periodically touches the side of the cup. Over the global, resonating waveform supported by the brain in its ongoing specification of this dynamic event, all these invariants—optical, acoustic, kinesthetic— will intrinsically be in sync, that is, already intrinsically "bound." What proposition (e.g., "The spoon is moving periodically") or set of propositions is going to capture these dynamics, reflecting, capturing perfectly, all this multimodal change, instant (state) by instant (state) by instant (state)?

The propositions are already only *abstractions*. They already rely on Peirce's "abduction," as we shall see, a fundamental form of inference neglected by AI (or mistakenly assumed to be handled), a form prior to both induction and deduction. The entire corpus of a linguistic text that a GPT processes, perhaps nightly, for achieving its statistical "embedding" of words is already one vast set of abstractions created by humanity. And abstractions mean nothing, they do not exist, without the concrete, without concrete experience.

Pylyshyn, we saw (Chapter 1) focused on "folding," noting that our understanding of folding depends on how folding is "organized," an organization apparently achieved in his database of propositions in 1973 and left problematic in 2002. "Folding," however, is an abstraction over concrete events of folding. Folds are made with towels, bedsheets, elbows, napkins, hexagonal number-forms in Penrose proofs, and at higher levels of abstraction (via analogy), poker hands, and debates. Folds do not exist without these concrete events, that is, without our dynamic images/experience of these events, in fact, as noted, without our resonant, "redintegrative wave" driving through our 4-D experience with its "stack" of folding events, the "wave" now defining an invariance (the abstraction—"folding), washing out the variants. This redintegrative process is the origin of the "compositional" elements (like folding, bending, napkins, elbows, etc.) that Fodor and Pylyshyn [Fodor1995] thought one of the two necessary components of human cognition (the other being "systematicity"). Fodor and Pylyshyn, along with cognitive science (and AI) simply assumed the existence of these compositional elements; these became the propositions in their databases. Then, paradoxically, they began looking for their origins within the very framework/device or metaphor (the computational) that in fact *can do nothing but simply assume them*, for the computational framework has no other resources to account for them.

THE PIAGETIAN BASE OF SYSTEMATICITY

The dynamical approach to compositionality was the essence of Piaget's approach. Consider his simple experiment on children aged 3–7 (*The Child's Conception of Movement and Speed*, [Piaget1946]). Here, three beads are strung on a wire which in turn can be fitted into a small cylindrical "tunnel." The beads are of different colors, but we will call them *A*, *B*, and *C*. The beads run into the tunnel and the tunnel semirotated from 1 to *N* times.

FIGURE 5.1. The tunnel-bead experiment.

A series of questions are asked, ranging from a simple, "What order will they come out?" after one semirotation (or half-turn), to the ultimate question on their order after any (*n*) number of half-turns. The child comes to a point of development where he can imagine the consequences of a 180° rotation which moves *ABC* to *CBA* and another 180° rotation which moves things back again to *ABC*, that is, an invariance of order under a 360° rotation. When now asked in which order would the beads come out when the tunnel is semirotated five (or 4, or 6, or 7, etc.) times, he evidences great difficulty. Some children appear to be exhausted after imagining three or possibly four semirotations, and they become lost when jumps are made from one number to another.

As Piaget describes it, the child, upon each half turn, endeavors to follow each turn and inversion of the bead order in every detail in his imagination. Given this effort, he only gradually manages accurately to predict the result of three, four, or five half-turns. Once this "game of visualizing the objects in alternation is set in train," as Piaget terms it, the child eventually perceives that upon each half-turn the order is changing. "Only the fact that up to this upper limit *the subject continues to rely on visualizing intuitively* and *therefore needs to image one by one the half-turn,* is proved because he is lost when a jump is made from one number of half-turns to any other" ([Piaget1946], p. 30, emphasis added).

After this gradual perception of a higher order invariant (the "oscillation of order") defined over events of semirotations, there comes a point then when the child can easily answer the ultimate question for the resultant order for any n-turns. Piaget's explanation, describing the "operational" character of thought, is foundational to his theory and its mathematical "group" operations.[1] "Operations, one might say, are nothing other than articulated intuitions rendered adaptable and completely reversible *since they are emptied of their visual content* and survive as pure intention ..." ([Piaget1946], p. 30). And he goes on to add that in other words, "... operations come into being in their pure state when there is *sufficient schematization*" ([Piaget1946], p. 30, emphasis added). So, instead of carrying out a concrete imaging of the tunnel's half-turn with the beads inside, that is, an *actual* representation, the child treats each half-turn as a *potential* representation, "like the outline for an experiment to be performed," but an experiment now not useful to do in detail, even mentally.

Thus, according to Piaget, operations, freed of their imaginable content, become infinitely *compositional*. This becomes the basis for forecasting the result of any n-turns, and it takes the child to about the age of seven. The operations become the generalization of actions performed through mental experiment. This is not simply abstract rules and symbols. It is not simple "rule learning." As we have

---

[1.] The rotations of a square about its center form a mathematical group of operations: (1) *closure* (all rotations, 90°, 180°, 270°, and so on, comprise a rotation within the group), (2) an *identity operation* (which leaves things unchanged, here a 360° rotation), (3) an *inverse operation* which undoes any rotation, e.g., a +90° and a -90°, (4) *associativity* (where, say, two rotations combine to make a third operation yet within the group, e.g., a 90° + 180° = 270° rotation).

seen, these "schematic" operations are built upon, and do not exist without, the dynamic figural transformations—*the dynamic imagery now employed in symbolic operations*—over which invariance emerges. They too are the result of a dynamical developmental trajectory incorporating these figural transformations, a trajectory which requires on average to the age of seven years.

It is worth noting here that it was precisely this form of developmental trajectory, in this case over the first two years, that results in COST (causality, object, space, time) and the ability to support explicit memory. It is precisely these developmental trajectories, crucial to the self-organizing system that is the brain, that the robotics theorist seems to believe can be simply ignored, somehow manufacturing the results in one fell swoop in his robots.

Current cognitive science has persuaded itself that the computer model of mind can handle Piaget and his "abstract" operations. At this point, it should be clear that this is badly mistaken. These operations are simply schematic transformations born of the concrete images of events—the exact thing neither AI, nor Pylyshyn, nor computer modelers know how to account for. Beneath Piaget's operations there resides Bergson's (and Gibson's) dynamical, resonating, 4-D "device."

Whether we deal with invariance over transformations, or change over transformations, we have entered the realm of time. Transformations are time extended. Cubes are rotating. Tunnels are half-turning. Flies are buzzing by. The Penrose hexagons are "folding." Wertheimer's line is breaking, rotating, and closing around a polygon. As we have seen, this time-extension cannot be conceived simply as an abstract series of "instants." To support intelligence, to support cognition, we require a device that supports time-extended transformations and invariance. In other words, for cognition we require consciousness; we need a memory that spans the "instants." We need a conscious device.

C.S. PEIRCE AND ABDUCTION

Many of the subjects we have discussed—redintegration, a model of perception, analogy, invariance laws, and transformations over an indivisible time—come together under C.S. Peirce (Figure 5.2)

FIGURE 5.2. Charles Sanders Peirce, 1839–1914, American philosopher, logician, mathematician, and scientist who is sometimes known as "the father of pragmatism."

and his discussion of *abduction*. Abduction, Peirce argued, is part of an intrinsically related triad—deduction, induction, and abduction. Without all three, we do not have intelligence. For Peirce, every observation that shapes the complex of ideas and judgments of intelligence begins with a *guess* or what he termed an abduction.

Erik Larson [Larson2021], who has written one of the great expositions of Peirce's thought, notes a passage from Peirce as a key starting point. In it, Peirce describes himself looking out his window in the spring and seeing an azalea in full bloom. Peirce notes immediately that this is only a description, a proposition, a sentence, a fact. But his perception of the azalea is far from a proposition, rather, it is an *image* which he makes intelligible in part by means of a statement of fact. The statement is abstract, the perception is concrete. This statement, this proposition, is an *abduction*, and is so for anything we see. Thus, Peirce would argue that, "… the whole fabric of our knowledge is one matted felt of pure hypothesis confirmed and refined by induction. *Not the smallest advance can be made in knowledge* beyond the state of vacant staring without making an abduction at every step" (Larson, p. 25, quoting Peirce, emphasis added).

For AI, this is not good; we are starting with perception—the image of the external world. Just like the coffee cup and stirring spoon, we must *see* those azaleas. The proposition, however, as Peirce notes, "I am seeing azaleas," just like "I am stirring the coffee," does not come close to the reality of the perception of the azaleas or the concrete event of stirring the coffee. We have already picked out something, isolated it, abstracted something from our experience, or, as we are going to discuss, decided on an aspect that has *relevance*.

We must note in Peirce's previous statement, the absolute centrality of abduction for Peirce: it is abduction that is responsible for the *advance of knowledge*. Any characterization otherwise is missing the point (and later we shall see just such a mischaracterization). Where does "abduction" fit with induction and deduction, particularly in the context of AI?

AI AND DEDUCTION

AI (taken in its initial phase as GOFAI or "Good Old-Fashioned AI") started using the deduction path. The schema for deduction is simple (Figure 5.3). Following this form, proof is guaranteed! But ... is there any *relevance*? One could have, following Larson's schemas: $A \Rightarrow B$: If it's winter, then the chickens will lay rocks. B: It's winter. A: Therefore, the chickens are laying rocks. This is a perfectly valid *modus ponens* argument, but the premise is false. Another case: $A \Rightarrow B$: If it's winter, then eagles will fly. B: It's winter. A: Therefore, eagles will fly. This is also valid, but the fact that its winter has nothing to do with eagles flying. Again, no relevance. And another: $A \Rightarrow B$: If it's Saturday, then people will go shopping. B: It's Saturday. A: People are going shopping. But there are numerous reasons other than it's Saturday why people could be shopping.

A => B If it's snowing, then the ground is white.

B It's snowing.

A The ground is white.

This is a classic deductive rule or form:
modus ponens.

FIGURE 5.3. Fundamental schema of deduction.

Obviously, the power of deduction depends on knowledge. This led GOFAI to massive attempts to create knowledge bases to guide deduction. One such example was Lenat's CYC project (Figure 5.4) [Lenat1983] (a project on which Larson worked for years). GOFAI and deductive efforts dominated into the 1990s, then came the Web and big data. GOFAI and knowledge bases were abandoned. Enter the reign of induction.

CYC project: Doug Lenat

This consisted of manual
entry into a DB of the
infinity of "facts" that
humans seem to know:

- Roads are flat.
- Cars drive on roads.
- Cars have wheels.
- The wheels can turn in different
  directions.
- Snow falls on roads.
- Snowplows remove the snow.
- And on, and on...

- This was a Sisyphean task, the DB
  massively unwieldy, with no end in sight.
  - (Erik Larson worked on this...)

FIGURE 5.4. Lenat's CYC project—for creating the encyclopedia of "facts" of which human knowledge is said to consist.

AI AND INDUCTION

Induction involves acquiring knowledge from experience, from observations, and generalizing from the observations. The primary mechanism for this is *enumeration* (Figure 5.5). Modern AI (including the GPTs) is based in this statistical framework, that is,

Enumeration:

- Big number of swans: they're always white.
- Therefore: All swans are white.

There could be a:

What if swans are white because of habitat?
It's a safe color given where they live.
Another habitat = black is good?

Another form (*statistical sampling*):
- 70% of farm chickens lay eggs daily.
- Therefore, with 10 chickens, we will get
  seven eggs today. .

FIGURE 5.5. Enumerative induction.

in the inductive framework. Of course, as Larson points out, we know the intrinsic problem: per Hume, we must believe that: "… instances of which we have no experience resemble those of which we have had experience" ([Hume1874], p. 390). There could be a black swan. What if swans are white because of their habitat? It (white) is a safe color given where they live. Might another habitat mean black is good?

As Larson notes in his review of the philosophical history of this subject, Hume's critique was related to causation: enumeration does not equal knowledge of cause. Induction does not have the resources to answer this. Nor does deduction—save partially at best. Hence AI theorists such as Judea Pearl [Pearl2000] are working on causal reasoning. AI is casting intelligence as detecting statistical regularities. But the problems have been noted: A net trained to classify images of "parked cars" on a training data set, will, on a test data set, classify as a "parked car" a car sitting half-submerged in the ocean, a car perched on top of a concrete highway divider, or a car "floating" in the air near the side of the road, but five feet above it. On these difficulties, the AI theorist, Joshua Bengio noted: "… [they] tend to learn statistical regularities in the dataset rather than higher-level abstract concepts" ([Marcus2019], p. 62). But our "abstract concept" is via our concrete experience of driving/parking cars—and our "imagination." One can imagine how the car got perched atop the concrete divider, one can do the visual transformations or "counterfactuals," and it did not get there by anything someone with the experience of parking understands as "parking."

The current AIs, particularly the GPTs, are enumerative induction engines, powerful, given the massive datasets for training, and taken well beyond what folks in the field once imagined such a process is capable of.

AI VERSUS ABDUCTION

When we seek to understand particular facts such as "Why is the swan black?" or "Why is the car on top of the concrete highway divider?" we enter the realm of abduction.

- Induction: from observations to regularities
- Abduction: from events to their causes

In Peirce's mind, abduction is associated with *surprise.* The everyday, ecological world is a constant stream of such. These surprises are also "clues" to their causes. Larson evokes hunters—the tracks, the broken twig, the matted grass, the sounds, and so on. (Figure 5.6)

Matted
grass

The fact, B, is observed. It is surprising!

If A were true, B would follow.

So, A could be true.

FIGURE 5.6. The schema of abduction/surprise.

The author is no deer hunter, but this is one "hunting" experience with instantaneous, complex abduction that is illustrative. It occurred in 1968 while in the US Marines in a reconnaissance battalion in the jungled hills of Vietnam. (For any reader looking for ecological examples, simply replace the recon team here with several Stone Age hunters.) The routine was to drop six men—the recon team— by helicopter into the middle of the jungle somewhere, stay in the area usually five days, avoid being detected, and of course chart the presence of the enemy, in this case the North Vietnamese Army (NVA). The hills were, in reality, mountains, roughly 550 meters in height or more. In the early evening of the first day of this recon patrol, we stopped on a finger leading up to a large mountain-hill in some low grass (about 3' tall) below the jungle covering the top of the larger hill further up the slope and ate some C-rations with the plan to move later to our night position. We moved out near dusk, quickly hitting taller grass and went barely several yards, and ... spots of matted grass—several of them! The (instant) *abduction:* Several NVA had been resting right by us, had gone the short ways further up the hill into the jungle trees to fetch a *much* larger unit there, and were soon coming down—for us! We instantly took a hard right to the other side of the finger, getting into (a bit of) cover. Within thirty seconds we heard a unit crashing down through the jungle, probably 100+, flashlights sweeping over our old eating spot and then starting to search over the larger area. Suffice it to say, we survived, but the instantaneous complexity of the inference and the instantaneity of action was, to the author, a striking aspect of mind.

A => B   If men were laying down, the
              grass is matted

B   The grass is matted.

A   Men were laying down.

A logical fallacy: Affirming the consequent.
There's ALL KINDS of reasons why the grass
could have been matted.
      Sasquatch
      Deer
      A python
      And on...

FIGURE 5.7. The schema of abduction and its problem (adapted from [Larson2021]).

So, there is a conjectural aspect to abduction, as we have just seen—a conjecture to plausible hypotheses. This leads to "logical trouble" (Figure 5.7), for it is the logical sin of *affirming the consequent.* Per Larson: "Viewing the logical form of abduction as a variant of bad deduction helps explain why it has been ignored, historically, in studies of reasoning, and why it has resisted mechanical methods ..." (Larson2021], p. 171).

Peirce's schemas (Figure 5.8) show why these forms cannot be reduced to one another. Again, as Larson notes, abduction cannot be another form of deduction because, "... its logical form is an egregious *deductive fallacy* ... it begins with *a conjecture* or a guess, which by definition might be wrong ..." ([Larson2021], p. 172).

The three are distinct, but intrinsically interdependent for intelligence. Induction is inadequate, and deduction is inadequate: AGI requires abduction.

The fact is, when we look at the characterization of abduction, we see a set of components only supportable by the Bergson-Gibson "device":

- causal reasoning

- imagination

Deduction

A => B	If men were laying down, the grass is matted.
A	Men were laying down.

B	Therefore the grass is matted.

Induction

A	We've observed millions of men laying down.
B	The grass is always matted.

A => B	Therefore when men lay down the grass is matted.

Abduction

A => B	If men were laying down, the grass is matted
B	The grass is matted.

A	Men were laying down. (Run!)

FIGURE 5.8. The three forms of inference (adapted from [Larson2021]).

- counterfactuals
- surprise
- relevance
- analogical reminding
- "storage" of all our concrete experience
- actual perception—*seeing* the coffee being stirred (i.e., experience)

Let's look briefly at some of these.

Abduction and Surprise

To remind again, the "specification" of direct perception is to a time-extent of the past transformation of the field: the fly, his wingbeats, the cup, the spoon's motion are in the past. This specification is enabled by the field's transformation being indivisible (not "instants" or "states"). We are talking then of 4-D being; our experience is not stored in the brain, but it is accessible—in its entirety. So, we come to a prime element of abduction—surprise.

For most of us, there are plenty of coffee stirrings in our 4-D being. Each event has a similar invariance structure, a structure driving the brain's modulated wave-pattern. A current event of coffee stirring resonates via this invariance-structure with these similarly structured events of past experience. The current event with its structure, as noted in Chapter 4, can be viewed as driving a wave through 4-D experience, all similar events resonating with it (Figure 5.9). Now introduce an *anomaly*, that is, some anomalous form of coffee stirring. The coffee liquid, for example, rises in a coherent column about an inch above the cup, then subsides, then repeats and repeats this pulsing, semi-geyser like performance.

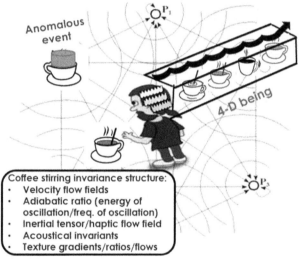

FIGURE 5.9. The coffee stirring, with its invariance structure, drives a wave (a resonance) through the 4-D extent of being.

Instant dissonance! Surprise! This we should note is equivalent to the frame problem. Let our being be a robot, stirring the coffee. How does the robot know this pulsing column of coffee is not an *expected feature* of the event? He must check his frame axioms, that massive list of propositions that is attempting to exhaustively list those things about and surrounding the event (there is also that problem of the context of the event) that do not change. The process for humans, at least as described here, is obviously entirely different— again, instant, *felt* dissonance, that is, an *intrinsic intentionality* (Cf. [Crockett1994]). And now the fun of abduction begins.

Abduction and the Source of Hypotheses

"What is going on with this event?" we ask. Now we need hypotheses. And where do these come from? We start with the fundamental retrieval operation: redintegration. The bulging/falling/rising coffee-event has a new structure:

- A rising mass
- Mysteriously constrained within a column
- Falling, rising again, rhythmically

What does the new event structure redintegrate? There is a different set of experienced events resonating:

- A piston driving up and down
- A glump of dough, with its yeast, rising
- A UFO, with tractor beam, drawing up a cow, then dropping it

And with each, there is a corresponding hypothesis:

- H_0: A cup with a piston-like bottom? (But how does the coffee stay constrained within a column?)
- H_0: Is there some sort of yeast in the cup?
- H_0: Some sort of force drawing the liquid up, like a UFO?

Redintegrating from 4-D experience via cues (or clues) is the start of the hypotheses, the conjectures. It is also *analogical reminding*.

Abduction and Analogical Reminding

For Hofstadter and Sander [Hofstadter2013], as we saw, *all* thought is founded on analogy, where "… each concept in our mind owes its existence to a long succession of analogies made unconsciously over many years, initially giving birth to the concept and continuing to enrich it over the course of our lifetime" (p. 3).

The mechanism of analogical reminding, however, eluded them. They saw it required "storage" of the totality of our experience, but how? We noted that they (not very seriously) considered and rejected events as stored (in the brain) on CDs of unindexed videos, instead reluctantly giving in to the concept that only abstractions of events, "gist," schemas, are "stored" despite the contradiction that this involves for accounting for the infinity of possible analogies.

This is the point for here: Peirce's "abduction" fully embraces the operation, problem, and centrality of analogy. And it rests on redintegration with its inherent aspect of analogical reminding.

Abduction, Imagination, and Counterfactuals

Imagination is central to coming up with hypotheses.

- a hidden piston?
- a tractor beam?

Again, for this, we must have solved *perception* in the first place—the origin of the image of the cup, that is, the origin of images, then of re-instantiating images from experience. Then we require a being embedded in the indivisible transformation of the universal field. We need the indivisible continuity of these visualized transformations; else we cannot perceive the globality of the transform—and the invariance preserved over it. Yes, we are back to Penrose's noncomputational thought.

Abduction and Relevance

John Vervaeke has been a tireless exponent of the need to solve the problem of *relevance*, to not simply give this away to the AIs: "But what hasn't happened … we haven't taken the emerging intelligence, the AI, and achieved real relevance realization by making it embodied, and we haven't joined the social/cultural—but we can" [Vervaeke2023].

In his long examination of the problem [Vervaeke2019], he comes to Newell and Simon's *Human Problem Solving* [Newell1972] as yet another of his many examples of the problem of relevance. We will use our own description of the problem here [Robbins1976]. Newell and Simon (henceforth, N&S) put problem solving in these terms:

- an initial state
- a goal state
- a heuristic procedure (their general problem solver [GPS]) which finds the sequence of steps to move from initial state to the goal state

Per Vervaeke, the brilliance of this heuristic is that N&S realized the combinatorial explosion problem. One only must think of the exponential number of possible sequences of moves emanating from a given chessboard position. The heuristic avoids the combinatorial explosion involved in considering all possible steps/paths. An N&S example is "The Monkey and the Bananas" problem. The situation is that the monkey is in a room, the bananas hang from the ceiling out of reach, and in the room is a box (which the monkey could move under the bananas, climb upon, and reach the bananas). There are a set of "relevant differences" given to the program and a set of operators (Figure 5.10).

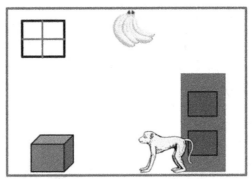

Relevant Differences:
- D1: The monkey's place
- D2: The box's place
- D3: The contents of the monkey's hand.

Operators:
1. Climb
2. Walk
3. Move box
4. Get bananas

FIGURE 5.10. The monkey and the bananas problem.

In N&S terms, all this is the *object language* in which the problem is framed. The program must discover the proper sequence of operators to reduce the differences successively to the point where the bananas are in the monkey's hand. In essence this a *proof* we can get from *A* to *B*, a proof procedure within a fixed theory (the object language) of the monkey's world. A fixed theory—for there is an indefinite set of other possible operators and other relevant differences (Figure 5.11).[2]

[2.] Narasimhan [Narasimhan1969] early on noted that the problem is *creating* the object language, else we only have a proof procedure carried out within a predefined framework.

Relevant Differences:

- D1: The monkey's place
- D2: The box's place
- D3: The contents of the monkey's hand
- D4: Monkey and door
- D5: Monkey and window
- D6: Two spots on the floor
- D7: Banana 1 and Banana 2
- D8: Banana and Monkey's tail
  And more…

Operators:

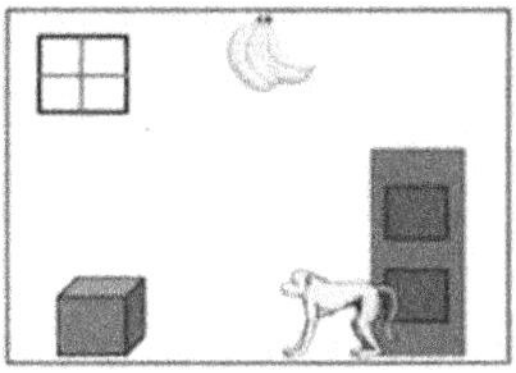

1. Climb
2. Walk
3. Move box
4. Get bananas
5. Twirl
6. Scratch nose
7. Paw on the floor
8. Summersault
9. Circle bananas 7 times
10. Blow very hard…
    And more…

FIGURE 5.11. Other possible operators and relevant differences.

This makes even this problem also an example of the combinatorial explosion problem. *Relevance* as Vervaeke would say, has been decided for the program. But there is more. The operator, "move box," is equally giving away a relevant feature of the world for the problem. Giving away the mobility of the box is roughly the same as giving Gutenberg the key to creating the printing press.

Consider Gutenberg's own account [Montmasson 1932], starting with his first letter to Frere Cordelier. His goal was to write in a "single instant," by "a single effort of my thought, everything that can be put on a large sheet of paper, lines, words, letters …" His first suggestion as to the means came when he noticed how playing cards were made, for the card images were carved on a block of wood, ink applied to the engraving, a thin sheet of paper placed over it, then rubbed, polished, and removed, leaving the image. This method had the advantage of being able to produce thousands of cards by repetition of a single operation and the multiplication of the blocks. Gutenberg wished to generalize this method, "for large pages of writing, for large leaves covered entirely on both sides, for whole books, for the first of the books, the Bible ..." (p. 272). But he did not see how. From his second letter:

> Every coin begins with a punch. The punch is a little rod of steel, one end of which is engraved with the shape of one letter, several letters, all the signs which are seen in relief on a coin. The punch is moistened and driven into a steel plate, which becomes a hollow or stamp. It is into these coin stamps, moistened in their turn, that are placed the little discs of gold, to be converted into coins by a powerful blow. ([Montmasson1932], p. 272)

From the fifth letter:

> I took part in the wine harvest. I watched the wine flowing, and going back from effect to cause, I studied the power of this press which nothing can resist. ([Montmasson1932], p. 273)

Continuing:

> To work then! God has revealed to me the secret that I demanded of Him... I have had a large quantity of lead brought to my house, and that is the pen with which I shall write. ([Montmasson1932], pp. 272–273)

At this point there occurred the relation to a seal:

> When you apply to the vellum of paper the seal of your community, everything has been said, everything is done, everything is there. Do you not see that you can repeat as many times as necessary the seal covered with signs and characters? ([Montmasson1932], p. 273)

And the whole process:

> One must strike, cast, make a form like the seal of your community, a mould such as that used for casting your pewter cups, letters in relief like those on your coins, and the punch for producing them like your foot when it multiplies its print. There is the Bible. ([Montmasson1932], pp. 273–274)

And the kicker (eighth letter):

> The letters are moveable. *The mobility of the letters* is the true treasure which I have been searching for along unknown roads. ([Montmasson1932], p. 274)

He has noticed the "repetitive" feature of card-making and the "multiplicative" feature of the card blocks. Being tied to the concrete transformations defining card making, they are very specifically defined. As they stand, these features have only their particular meaning in their particular case. Notice, it is not even clear to Gutenberg that these are particularly important features. The "features" involved in coin-making—the rigidity and force provided by the steel punch are likewise tied to the particular event. These events were fused in Gutenberg's mind with yet a third event—the wine pressing.

There are invariants defined over all three, particularly the application of great force and the repetitive application of the force. These features can now approach the abstract generality which the feature matching framework envisions [Gentner1983], but they do so only by virtue of their definition across concrete events. Gutenberg was *defining the features* of the world in which he would solve the problem. *He* was creating the object language. The features were being defined through analogy—he was seeing analogies to seals, to molds, to stamping coins.

The analogy defines the features. The "features" are what we term the "relevant." We do not start with, we are not given, the features (cf. [Robbins1976]). As we start as infants, just as later when we become inventors, the brain is initially confronted with a world in which it must define the features. It is the problem of the definition of features that must be solved. We could replace the coffee stirrings in Figure 5.9 with wine pressing, card engraving, coin stamping, seal making. This same redintegrative basis—invariance structures, 4-D being—is operating in Gutenberg's case.

Vervaeke noted this: all inference presupposes *relevance realization* ([Vervaeke2019], at 38.00). This was exactly Peirce's thesis: Deduction and induction both require abduction—abduction is at their base.

Abduction and Causal Reasoning

"Causal Reasoning" has clearly been involved in what we have been describing here. Why is that coffee-liquid pulsing up and down? What is the *cause*? We are not saying there is a detailed understanding of how all this works within the general framework of the Bergson-Gibson "device." The complexity of what we have just seen in the context of Piaget's COST and his "operations," the developmental trajectory of the ability to learn the "simple" odd-even rule (invariant) for tunnel rotations should disabuse us of any notion that all this is easily explained. But we require these basic elements:

- The brain as a modulated reconstructive wave passing through a holographic field, specific to a source—by this process an image
- The transformation of the field—indivisible, continuous
- The self as a 4-D being, "trailing" all experience
- The events of experience structured by laws of invariance

The previous aspects comprise the core reason why abduction will not yield to the neural nets. The last bullet, however, the invariance laws, is at the heart of the problem of causal reasoning. The laws of physics are invariance laws, for example, $E = mc^2$, $F = -kx$, $F = ma$, and so on. The law, $F = -kx$, is expressing, for a range of stretching transformations on springs over a distance (x), an invariant effect of a restorative force (F). This involves us in the question: is this a "causal" law, or is it just what physics must actually and only express, namely, an *invariance law*? (Cf. Woodward [Woodward2000], [Woodward2001].) Taking the spoon in coffee stirring at a higher level of abstraction, it represents an invariant presence in the event of an object with rigidity, sufficient width (a toothpick is rather inefficient), grasp-ability, all sufficient to apply a force to move the liquid medium given the liquid's resistance. We understand the spoon as "causal" in producing this motion, but this is along with a slew of other "causes"—the hand-arm and its muscles, our intention, the food eaten to fuel the hand-arm, and so on. The spoon at least is part of the invariance structure of this event, as is the piston in the anomalous "pumping/geyser" coffee event about which we are hypothesizing causes. The bottom line here: for "causal reasoning," one must have a "memory" that supports the 4-D, time-extended invariance structures defined across events.

Abduction and the Problem of the Frame

The question was earlier noted, how does the robot recognize something in the event as unexpected? The *context* takes the problem beyond the narrow frame of just the coffee stirring event. What if, while stirring the coffee in the morning, five turkeys walk by the window? Is this anomalous with respect to coffee stirring, an unexpected change? It is entirely expected in the context of a minifarm, but this is, again, resonant with the minifarm experience. The garbage truck comes by; it picks up the trash, making a great noise. Unexpected change while coffee stirring? Not really—a part of the 4-D experience of the minifarmer.

The problem of these "frames," in effect our vast "stored" experience, is reflected upon by Hofstadter and Sander [Hofstadter2013] in discussing AI language translation problems (pp. 367–383). They take an obituary (in French) for Francois Sagan, for which we'll go directly to the Google 2019 translation:

> Sometimes, the success was not at the rendezvous. We may think very hard about it, but the right number does not necessarily come out. Sagan took playwright's failures as casino setbacks, with respect for the whims of the bank and heaven. You must lose a little, to better enjoy the next day's win. That she did not see her "recover" in a few quarters of an hour the losses of a whole night can understand how joyful it is to taunt fate. (Google translate, 2019)

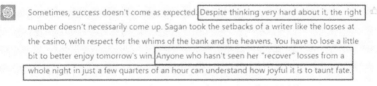

FIGURE 5.12. ChatGPT translation (April 5, 2023).

A ChatGPT translation (April 5, 2023) is shown in Figure 5.12. Hofstadter and Sander provide a Google 2009 translation along with their own *human* translation. One can see the improvement in the AI, and the remaining areas of difficulty. The "frame" is the world of casinos, where the "bank" is the house, and of a being (Sagan) who loved gambling. This frame, well, never quite gets framed:

- French : Pour les caprices de la banque et du ciel.
- Google, 2019 : For the whims of the bank and the sky.
- ChatGPT, 2023 : For the whims of the bank and the heavens.
- Human (H&S) : The whims of the house and of divine fate.

"Ciel" can be: sky, heaven, blueness, air, midair, a vault, canopy, ceiling, clouds, the atmosphere, space. The frame will determine the selection. Google (then, in 2019) did not "understand" the frame; ChatGPT, with much vaster data, *appears* to understand this, but not really—the human experience is just not there; it still struggles with the vast difference between "thinking very hard" about something and this feature of the frame: the wishing like crazy that the dice roll comes up to one's desires.

- French: On a beau y penser très fort, le bon numéro ne sort pas forcément.
- Google, 2019: We think very hard, the right number does not necessarily come out.
- ChatGPT, 2023: Despite thinking very hard about it, the right number does not necessarily come up.
- Human (H&S): No matter how hard she wished for it, the dice simply wouldn't come up her way.

Finally, there is that last sentence of the obituary, in the H&S translation, "And if you never saw her win back a whole night's losses in well under an hour, you just can't have any idea of the glee she took in laughing at the face of destiny." This holds some French syntactic subtlety and is only translated correctly if one understands the human-gambling frame, and well, let us say this sentence is still undefeated (as of this date).

This is not to say there will be no more improvement. The transformer algorithm, one is almost compelled to say, was an assault on the pronoun reference problem, for example, sentences like:

- The cat ate the rat because it was hungry. [Who was hungry?]
- The chicken didn't cross the street because it was too tired. [What was too tired?]

As noted, the weighting (clustering) of words in the embedding process and the subsequent "attention" algorithm greatly alleviated this difficulty. But it is just a way to avoid actual knowledge of the frame or frames, that is, to avoid having a model that truly embraces a being with perception and experience. Claims that the brain is actually doing things like the GPTs, just not as efficiently, carry zero credibility.[3]

UNDERESTIMATING ABDUCTION

Roli, Jaeger, and Kaufman argue [Roli2022], very much along the lines of the mousetrap problem [Robbins2012], that there are deep implications in the fact that the functions of something like a pencil or screwdriver cannot be exhaustively specified, that there is no definite list. Roli et al. note that general intelligence involves "... situational reasoning, taking perspectives, choosing goals" (p. 1) and they relate this to the Gibsonian concept of affordances, that is, identifying and exploiting new affordances, where a general example of such is the use of an object in the hands of an agent. Thus, they argue, "... *it is impossible to predefine a list of such uses*. Therefore, they cannot be treated algorithmically ... This implies that true AGI is not achievable in the current algorithmic frame of AI research" ([Roli2022], p. 1).

There is a problematic feature of their argument that is of interest here before we go more deeply into their exposition of the problem, for they wish to note what they perceive as a neglected issue in the discussion of AGI, namely, "... for a long time, we have believed that coming to know the world is a matter of induction, deduction, and abduction (see, e.g., [Hartshorne1958], [Mill1963], [Ladyman2001], [Hume2003], [Okasha2016], [Kennedy2018]. Here, we show that this is not enough" (p. 7).

The significance of the fact that AI has never implemented abduction seems to have escaped them.[4] As Larson points out

---

[3.] For analysis of the close relation of Bergson to Peirce, see [Bankov2000].
[4.] Given Roli et al. place critical emphasis on the *semiosis* of Peirce, it might be considered a bit contradictory to be downgrading his abduction, something that he saw as central to intelligence.

([Larson2021], pp. 167–168), there once was an abductive logic programming (ALP) effort with these attributes:

- Abduction was treated as a form of *deductive* inference.
- It derives from a *knowledge base* to the truth of $A \Rightarrow B$, where B is the observation, A is the explanation thereof.
- Problem: this eliminated the conjectural basis of abduction; thus, *no inferential power is gained*.
- Hence, ALP was abandoned.

So, in the ALP effort, the researchers were basically trying to *load a database* with the myriads of $A \Rightarrow B$s, that is, they were trying, again, to *fix* the world, to create an object language, to hold it steady, motionless, essentially to *remove surprise* (for all events are already in the knowledge database), thus eliminate conjecture. But this is exactly what one cannot do and still have abduction.

Roli et al. place the issue in the context of diagnosing problems with automobiles, arguing that there is no unique (and fixed) decomposition of an automobile (into parts) that rests handily around for use by the problem solver. Looked at as a vehicle to get us somewhere, there is one functional decomposition; looked at as a device for frying eggs (using the hot engine block), there is a quite different functional decomposition. We might note here, for fun, that this was exactly Pirsig's point in his description of motorcycle decomposition in his *Zen and the Art of Motorcycle Maintenance* [Pirsig1974]), for all decomposition is subjective, and the decomposition given in your boring, technical parts manual for the motorcycle may be useless for your particular problem.[5]

Roli et al. state: "Abduction is a differential diagnosis from a *prestated set of conditions* and possibilities that articulated to carry out what we 'see the system as doing or being'" (p. 8, emphasis added). But they note, there is no unique decomposition; the number of decompositions is indefinite. Therefore, computer programs, using this form of reasoning "... cannot reveal novel features of the world not already in the ontology of the program" ([Roli2022], p. 8), where

---

[5.] For an examination of the unnoticed and very deep relation of Pirsig's *Zen and the Art* to the philosophy of Bergson, see [Robbins2018].

"ontology" here should be considered precisely the predefined "object language" we have already discussed.

The "prestated conditions" statement of Roli et al. is the equivalent of the second bullet in the description of the failed ALP attempt—the "knowledge database" with a predefined set of conditionals. But relying on a set of prestated conditions is precisely what abduction does not do, and precisely why it cannot be implemented by AI. In fact, abduction, when understood as consisting of the components discussed previously (redintegration, analogical reminding, imagination …) is the very basis for revealing those "novel features." Given our discussion here on Peirce's vision of abduction (versus Roli et al. description of abduction), it is safe to say there is a severe underestimate out there of what abduction really involves.

ROBOTS VERSUS ECOLOGICAL INTELLIGENCE

Roli et al. address a question that is critical to this book, and we will key off of it here. They note the need to go beyond algorithms that run only within some stationary computing environment, for in this environment, its purely virtual nature implies that all features within must, by definition, be formally predefined. They put it as such (and note the echoing of Peirce re surprise): "The real world is not like that. We have argued … that *our world is full of surprises* that cannot be entirely formalized, since not all future possibilities can be prestated" (p. 8, "surprises" emphasis added). So, they ask if an AI agent working in the concrete, ecological world, "… could identify and leverage affordances when it encounters them. In other words, does our argument apply to *embodied Turing Machines*, such as robots …" (p. 8).

Or put in the context of this book, can a robot achieve ecological general intelligence? The Roli et al. answer is, "No." Their route to this "no" has important points, but it is missing the larger "frame" in which it should be placed. Firstly, their argument involves delineating what they argue are key differences between an organism and a machine. This begins with an organism having goals, setting its own goals (as opposed to a machine), then the establishment of boundaries required to support the autopoiesis of Maturana and

Varella [Maturana1980] and they note that when the "boundary processes and constraints" discussed by Moreno and Mossio [Moreno2015] have been integrated into this closure of constraints, the organism attains a new level of autonomy: *interactive autonomy*. "It has now become a fully-fledged *organismal agent, able to perceive its environment* and select from a repertoire of alternative actions ..." (p. 5, emphasis on "to perceive ..." added).

Further on they will say, "This 'coming to know the world' is what makes evolutionary expansion of our ontologies possible. It goes beyond induction, deduction, and abduction. Organisms can do it, but universal Turing machines cannot" (p. 8). But, stopping here briefly to comment, autopoiesis is not a theory of *perception*; the "interactive autonomy" of Moreno and Mossio is not a theory of perception; the ability they claim "to perceive its environment" has been born via a magic wand of "processes and closure of constraints;" there is nothing in this account that is explaining the origin of the image of the external world. The "device" Roli et al. implicitly have in mind as comprising a perceiving organism, and consequently, we shall see, aspects of their concept of the nature of affordances, is not up to the task. *Perception*, however, as we have seen, is the very beginning of abduction. Of course, without a theory of perception—how Peirce sees those azaleas—one cannot understand abduction and will downplay it, but everything we have subsumed under abduction is needed for the not-a-robot, organismic alternative they implicitly envision.

Focusing directly on the embodied robot problem, Roli et al. argue that the robot, even though acting on its surrounding environment via sensory receptors and effectors, will still crash against both the symbol grounding problem and the frame problem, the latter requiring that the robot find what is *relevant* to achieving its goals (e.g., are the turkeys relevant, the garbage truck?), this relevance being something that cannot be formalized. As an example, they have us imagine a robot whose task is to open coconuts and whose only available tool is an engine block—a block which is currently perceived (categorized) as a paper weight. They ask if it can perceive the affordance(s) implied by the engine block? The robot might acquire experience via random moves; it might hit the engine block, discovering that the block has the property of being "hard and sharp," which (from an omniscient observer's perspective) is useful for cracking the nut. But given this property is not in the pre-defined object language, "… how does the robot know that it needs to look for this property in the objects of its environment?" (p. 9).

Yes, this is exactly the monkey's problem of knowing that the "mobility" of the box is critical, or the critical mobility of type for Gutenberg. They go on to note that by the same random moves, the robot might move the engine block or tip it, but how does the robot understand that "hard and sharp" are useful, but in this case "move to the left" is not? In other words, again, there is no list of pre-defined operators like "move the box" as a predefined object language. So, they add: "Furthermore, if the coconut is lying beside the engine block, tipping it over may lead to the nut being cracked as well. How can the robot connect several coordinated causal features to achieve its goal, if none of them can be deduced from the others? The answer is it cannot" ([Roli2022], p. 9).

There is no way for the robot to know, they argue, that it is improving over the incremental steps of the search. Once an affordance is identified, new ones emerge, but how can the robot know that it is accumulating successes until it chances upon the final achievement? "There is no function optimization to be performed over such a sequence of steps, no landscape to search by exploiting gradients …" (p. 9). Again, we are in a space of possibilities that cannot be predefined. They note that the only way to achieve the robot's ultimate goal is via the preprogrammed ontology (the object language, the operators) that allows for multistep inferences, for embodied or not embodied, the algorithm operates deductively. And importantly for the future-envisioned GPT-driven robots: "But if the opportunity to crack open coconuts on the engine block has been predefined, *then it does not really count as discovering a new causal property*. It does not count as exploiting a novel affordance" (p. 9., emphasis added).

The "preprogrammed ontology" is the object language and note that Roli et al. are envisioning a robot *acting randomly*, "hoping" to stumble upon a solution. If the robot is indeed attempting to simulate an organism, acting ecologically in its environment via its "experience" of the environment, but where this "experience" is just the stored events-as-features of Moscovitch et al., or even the unindexed videos Hofstadter and Sander contemplated with little seriousness, this argument of Roli et al. holds well. If, however, the robot is being driven by ChatGPT (better, some GPT-[n] down the road) and therefore using its distillation of masses of text representing human experience, where, as in the mousetrap task, the robot is essentially being given the query, "How can you use the engine block to crack the coconut?" it will come up with an instruction set that works, with the instructions even driving the robotic actions. But then the robot has fallen precisely into their critique—it has used predefined properties of the world and this "… does not really count as discovering a new causal property."

The argument is good; again, the frame in which the argument is placed is insufficient. The (non-GPT) robot would require the 4-D experience and redintegrative, content-addressing, analogical-reminding memory mechanism, and counter-factual imagination we have earlier described. It needs abduction. The time-framework of the temporal metaphysic is also required, for this thought process occurs as indivisible transformations allowing the invariance to be registered. It is a *building* time or flow, the moments interpenetrating, permeating, the present reflecting previous history. Roli et al. (along with Maturana and Varela) *assume* (unaware) this aspect of the temporal metaphysic, often noting how experience must build in time, for example, speaking of the machine: "There are no actions that emanate from the *historicity* of internal organization" ([Roli2022], p. 5, emphasis added). An organism is intrinsically embedded in the indivisible transformation of the universal field; the abstract operations of a machine are not; they have no relation to this indivisible, melodic transformation. Of course, perception is the start of all this—an actual image of the external world which simultaneously *is* that "stored" experience. But perception is *virtual action*—the display of possible action, and "affordance" is just another term for this—perception as the display of possible action, and at a *scale of time*. Roli et al. view affordances as falling within the standard view of the perception-action cycle—(see, then act on/ change the world, then see the result, then act again). Perceiving affordances—how the spoon might be used as a catapult-like pea-launcher—is really being thought of here as something in the mental sphere, like perceiving the odd-even rule for beads and tunnel-turns.

But no, perception is already suffused with affordances, with virtual action, and the holographic reconstructive process involved in these indexes the profound difference between machine and organism.

CONSCIOUSNESS, COGNITION, AND MOUSETRAPS

The Roli et al. argument, with its focus on the problem of (or non-effectivity of) random action, if one looks more closely, is in effect a critique of neo-Darwinian evolution. As argued elsewhere [Robbins2012] [Robbins2022], if taken in a larger frame, it indicates that evolution, at least as in the neo-Darwinian mode, cannot proceed via random chance combined with AI-like algorithmic processes, but must proceed via an entirely different form of (cosmic-in-scale) "device."[6]

Michael Behe started off the controversy, ostensibly in the evolutionary sphere, challenging the possibility of what can be termed an "algorithmic approach" to design espoused by evolution theory [Behe1996]. In reality, and why of interest here, the controversy sits squarely in the realm of *cognition*, and equally, in the realm of *commonsense knowledge.* Although Behe dealt heavily in the biochemical realm, he placed the problem initially in the intuitive context of a mousetrap. The (standard) mousetrap consists of several parts (Figure 5.13). As a functioning whole, he argued, the trap is "irreducibly complex." For the device to work as designed, all the parts must be present and organized correctly, else it does not function.

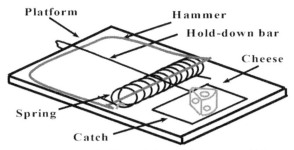

FIGURE 5.13. Mousetrap—standard issue. (Adapted from McDonald [McDonald2000]).

---

6. We would argue [Robbins2022] that the recent theoretical trend toward "self-organization" is not the answer in this case either.

The urge is to break the problem of instantiating this design into simpler components and steps—evolving the separate, smaller parts. Natural selection (a random process) buys nothing here, Behe argued. Natural selection picks some feature or form or component to continue because it happens to have been proven useful for survival. Evolving a single part (component) which by itself has no survival value is impossible by definition—impossible, that is, by the definition of the role and function of natural selection. But even if by chance the parts evolved simultaneously, there remains the enormous problem of organization of the parts. How does this happen randomly? Each part must be oriented precisely spatially, fitted with the rest, fastened down in place, and so on. There are enormous "degrees of freedom" here—ways the parts can rotate, translate, and move around in space—which drive the odds against randomness to enormous proportions.

Evolutionists initially attacked this "irreducible complexity" argument on the basis that one could easily show there are simpler, less complex traps, the implication being that therefore the improbabilities have been lowered. An argument, often cited as though it were a definitive critique, was provided by McDonald to demonstrate how the mousetrap could have simpler instantiations [McDonald2000]. His caveat is that this is not an analogy for evolution per se, but the argument *is* taken as a critique of Behe. Working backwards from the normal, "standard issue" trap, McDonald gradually simplified the trap, producing four "predecessor" traps of decreasing complexity. Behe argued, however, it is not that simpler mousetraps do not exist. The question is *progression*—the actual mechanism of movement from trap A to trap B to trap C. If McDonald is taken as a defense of evolution, Behe produces a strong counter argument [Behe2000]. Starting with McDonald's first and least complex trap (Figure 5.14, left) in the "sort of evolving" series, he examined the steps needed for McDonald to arrive at the second, more complex trap (Figure 5.14, right). The first (or single piece) trap has one arm, under tension, propped up on the other arm. When jiggled, the arm is released and comes down, pinning the mouse's paw. It is a functional trap.

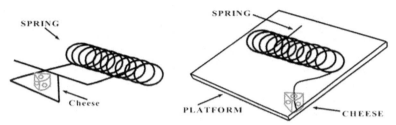

FIGURE 5.14. Traps 1 and 2 of McDonald's first series. (Adapted from McDonald [McDonald2000]).

The second trap has a spring and a platform. One of the extended arms stands under tension at the very edge of the platform. If jiggled, it comes down, hopefully pinning some appendage of the mouse. To arrive at the second, functional trap, the following transformations are needed:

1. Bend the arm that has one bend through 90 degrees, so the end is perpendicular to the axis of the spring and points toward the platform.

2. Bend the other arm through 180 degrees so the first segment is pointing opposite to its original direction.

3. Shorten one arm so its length is less than the distance from the top of the platform to the floor.

4. Introduce the platform with staples. These have an extremely narrow tolerance in their positioning, for the spring arm must be on the precise edge of the platform, else the trap won't function.

All of this must be accomplished before the second trap will function—an intermediate but nonfunctional (useless) stage cannot be "selected." This complicated transition is a sequence of steps that must occur *coherently*. With each step required, we decrease the probability of random occurrence exponentially. The "four" steps in the mousetrap transition are an arbitrary number, and indeed could be decomposed to many further sub-steps, especially when dealing with the level of random mutations.

Each of the subsequent trap transitions (2⟹3, 3⟹4, 4⟹5) proved subject to the same argument. McDonald then produced a more refined series of traps [McDonald2002]. He argued that the point was made that a complicated device can be built up by adding or modifying one part at a time, each time improving the efficiency of the device. Yet there are still problematic transformations between many of his steps—it is hardly "one part at a time." For example, in the second series, the transition between a simpler spring trap (Figure 5.15, Trap Five) and one now employing a hold-down bar (Figure 5.15, Trap Six) is a visual statement of the difficulty of the problem. Even if the simpler trap were to become a biologically based analog—a largish "mouse-catcher beetle" —sprouting six legs and a digestive system for the mice it catches, the environmental events and/or mutations which take it to the next step (as in Trap Six) would be a challenge to define.

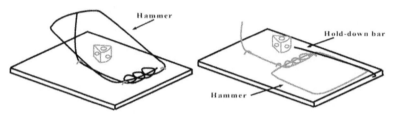

FIGURE 5.15. Trap Five (left) and Trap Six (right) of McDonald's second series. (Adapted from McDonald [McDonald2002]).

If we step back for a moment, we ask, "What is driving these transitions, say, from Trap 1 to Trap 2, or from Trap 5 to Trap 6?" Clearly, given the number of steps that could be involved, especially in the biological realm of molecules and cells, it cannot be random bumbling. The force of Roli et al.'s argument with respect to the robot comes home. How is the goal, say, to get to Trap 2, held in mind as a simultaneity while carrying out these steps? Even worse if the goal—Trap 2—is not even yet known? Perhaps more simply, how did McDonald arrive at Trap 1 (of the first series)?

We can argue, just as in searching for hypotheses explaining the rising/falling coffee column, that it was via abduction. For example, we see that the hammer-arm of the "standard issue" trap provides a downward, crushing force. How can this be done more simply? What are examples of crushing/pinning forces? Redintegrate such events, probe experience: perhaps a boot coming down, pinning a foot. Now what transformation of the hammer-arm would achieve such a force, and more simply? Why were the staples introduced? Because in imagination (counterfactually, causally) our event-experience shows a force is required for leverage against the pinning action of the wire. And so on. For this we have invoked 4-D experience-memory and the resonant, redintegrative "device." With this, there is the plethora of transformations—bending wires, stapling, twisting, shortening, lengthening, fastening. These are transformations, over time, just as in the Penrose cubes that, at minimum, leave something invariant—the length of the wire, the strength of the material. How does a robot, in actuality a Turing machine operating in its "time" of discrete states, perceive this? One requires a "device" operating in an indivisible time.

The most apparently decisive evolutionary argument that arose around this subject is that indeed biological "parts" exist that in themselves are independently functional. In essence, then, evolution has available to it pools of independently functional components from which to select, and from which to build various larger functioning wholes. Kevin Miller [Miller2004] considered this the finding of Melendez-Hevia et al. (1996) in the realm of the Krebs cycle [Mel1996]. Miller applies this logic to the mousetrap. Each component can be conceived to be an independently functional part. For example, the hold-down bar can serve as a "toothpick," the platform as "kindling," three of the components can work together as a "tie clip" (platform, spring, and hammer), and so on.[7]

But of course, this is simply falling into the other piece of the critique: there is no preset, predefined, exhaustively specified list of functions for these "parts," like pencils, rubber bands, box corners, and so on. These are again functions of the transformations, the "set" of transformations also being an open-ended list, and the transformations again occurring, necessarily, over an indivisible flow for the invariants involved to be registered.

[7.] Miller, it should be noted, rather badly mischaracterizes Behe's argument, somehow construing Behe as invoking a pool of parts where, "each part has no function" [Miller2006]. What sense a set of functionless parts could possibly make, or why someone would envision such, is hard to grasp, but the fact is we are dealing with a "pool" (indefinite) of parts whose functions are indefinite, where new functions emerge under transformations, and this latter view is the one consonant with Behe's argument. The problem is how we create (design) a fully functional device from a set of parts with entirely different functions, or, equivalently, transform functioning organism A to (functioning) organism B with one coherent transformation set.

Evolution's expositors have often tended to be oblivious to depth of the problem, which in truth is no more and no less than AI's problem of commonsense knowledge; they blow right by it. Richard Dawkins (*The God Delusion*), approvingly references Miller's concept, arguing that a mechanism used by parasitic bacteria for pumping toxic substances through cell walls is "similar enough" to a component in the bacterial flagellum—the little whip-like motor that moves the bacteria around—that evolution must have simply "commandeered" this component. For "commandeer," Dennett (*Darwin's Dangerous Idea*) substitutes the term "generate and test," holding, with no explication, that evolution simply "generates" new devices such as flagellar motors (or mousetrap #5) to test them out. In lieu of "commandeer," Shermer (*Why Darwin Matters*) confidently employs the term "co-opt," as in evolution "co-opts" features to use for another purpose. For "commandeer," Eugenie Scott (*Evolution vs. Creationism*) uses "borrowing and swapping." Kevin Miller himself simply uses "mix and matching" saying, "… it is to be expected that the opportunism of evolutionary processes would mix and match proteins to produce new and novel functions" [Miller2004]. Shermer [Shermer2007] notes approvingly Darwin's own statement, glossing over of the same problem in his concept of "exaptation":

> On the same principle, if a man were to make a machine for some special purpose, but were to use old wheels, springs, and pulleys, only slightly altered, the whole machine, with all its parts, might be said to be specially contrived for that purpose. Thus, throughout nature almost every part of each living being has probably served, *in a slightly modified condition*, for diverse purposes, and has acted in the living machinery of many ancient and distinct specific forms. ([Darwin1877], pp. 283–284, emphasis added).

"Slightly modifying" the parts for entirely new functions (or in other terms, perceiving the new affordances!) is precisely the problem. So, all this would seem to tend to the vision that, just as the brain, the universe, as a "mechanism" for evolution of forms, beings, "devices"—beetles, butterflies, brontosaurs—is indeed quite a different form of "device," something far from a Turing Machine. This is of course a massive question, but its overtones are going to be present as we examine the problem of voluntary action.

REFERENCES

[Bankov2000] Bankov, K. "Intellectual effort and linguistic work: Semiotic and Hermeneutic aspects of the philosophy of Bergson," *Acta Semiotica Fennica IX,* International Semiotics Institute at Imatra, 2000. Available online: https://helda.helsinki.fi/server/api/core/bitstreams/50d16c7a-1ad3-4842-a497-cef2d20339b0/content

[Behe1996] Behe, M. *Darwin's Black Box: The Biochemical Challenge to Evolution.* New York: The Free Press, 1996.

[Behe2000] Behe, M. "A Mousetrap defended: Response to Critics." http://www.arn.org/docs/behe.mb_mousetrapdefended.htmhttps://www.discovery.org/a/446/[Crockett1994] Crockett, L. J. *The Turing Test and the Frame Problem.* Norwood, N.J., Ablex, 1994.

[Darwin1877] Darwin, C. *The Origin of Species by Means of Natural Selection.* London: Murray (2nd edition), 1877.

[Fodor1995] Fodor, J. & Pylyshyn, Z. "Connectionism and cognitive architecture," in C. MacDonald & G. MacDonald (Eds.), *Connectionism: Debates on Psychological Explanation.* Oxford: Basil Blackwell, 1995.

[Gentner1983] Gentner, D. "Structure-mapping: A theoretical framework for analogy," *Cognitive Science* (1983): 7(2), 155–170.

[Hartshorne1958] Hartshorne, C., & Weiss, P. *Collected Papers of Charles Sanders Peirce.* Boston: Belknap Press, 1958.

[Hofstadter2013] Hofstadter, D. & Sander, E. *Surfaces and Essences: Analogy as the Fuel and Fire of Thinking.* New York: Basic Books, 2013.

[Hume2003] Hume, D. *A Treatise of Human Nature.* Chelmsford, MA: Courier Corporation, 2003.

[Hume1874] Hume, D. *A Treatise of Human Nature.* London, UK: Longmans, Green and Co., 1874.

[Kennedy2018] Kennedy, B., & Thornberg, R. "Deduction, induction, and abduction," in *The SAGE Handbook of Qualitative Data Collection* (49–64). London: SAGE, 2018.

[Ladyman2001] Ladyman, J. *Understanding Philosophy of Science.* London: Routledge, 2001.

[Larson2021] Larson, E. J. *The Myth of Artificial Intelligence: Why Computers Can't Think the Way We Do*. Cambridge, MA: Belknap Press.

[Lenat1983] Lenat, Douglas B, Borning, A., McDonald, D., Taylor, C., and Weyer, S. "Knoesphere: Building expert systems with encyclopedic knowledge." *Proceedings of the Eighth International Joint Conference on Artificial Intelligence—Volume 1. IJCAI'83* (1983): *1*, 167–169.

[Marcus2019] Marcus, G., & Davis, E. *Rebooting AI: Building artificial intelligence we can trust*, New York: Vintage, 2019.

[Maturana1980] Maturana, H., and Varela, F. J. *Autopoiesis and Cognition: The Realization of the Living*. Dordrecht: Springer, 1980.

[McDonald2000] McDonald, J. (2000). "A Reducibly Complex Mousetrap." https://udel.edu/~mcdonald/mousetrap.html https://udel.edu/~mcdonald/oldmousetrap.html

[McDonald2002] McDonald, J. (2002). "A Reducibly Complex Mousetrap." https://udel.edu/~mcdonald/mousetrap.html

[Mel1996] Melendez-Hevia, Waddell, & Cascante, M. "The puzzle of the Krebs citric acid cycle: Assembling the pieces of chemically feasible reactions, and opportunism in the design of metabolic pathways during evolution," *Journal of Molecular Evolution* (1996): 43(3), 293–303.

[Mill1963] Mill, J. *Collected Works*. Toronto, ON: University of Toronto Press, 1963.

[Miller2004] Miller, Kevin. "The flagellum unspun: The collapse of irreducible complexity," in M. Ruse & W. Dembski (Eds.), *Debating Design*, Cambridge UK: Cambridge University Press, 2004.

[Miller2006] Miller, K. "The Collapse of Intelligent Design: Will the Next Monkey Trial be in Ohio?" March 17, 2008, YouTube video. https://www.youtube.com/watch?v=Ohd5uqzlwsU

[Montmasson1932] Montmasson, J. *Invention and the Unconscious*. New York: Harcourt-Brace, 1932.

[Moreno2015] Moreno, A., and Mossio, M. *Biological Autonomy*. Dordrecht: Springer, 2015.

[Narasimhan1969] Narasimhan, R. "Intelligence and artificial intelligence," *Computer Studies in the Humanities and Verbal Behavior* (1969): March, 24–34.

[Newell1972] Newell, A., & Simon, H. *Human Problem Solving*. Englewood Cliffs, NJ: Prentice-Hall, 1972.

[Okasha2016] Okasha, S. *Philosophy of Science: A Very Short Introduction, 2nd Ed.* Oxford: Oxford University Press, 2016.

[Pearl2000] Pearl, J. *Causality: Models, reasoning, and inference.* Cambridge, UK: Cambridge University Press, 2000.

[Piaget1946] Piaget, J. *The Child's Conception of Movement and Speed.* New York: Ballentine, 1946.

[Pirsig1974] Pirsig, R. *Zen and the Art of Motorcycle Maintenance.* New York: Bantam Books, 1974.

[Pylyshyn1973] Pylyshyn, Z. "What the mind's eye tells the mind's brain: a critique of mental imagery," *Psychological Bulletin* (1973): 80(1), 1–24.

[Robbins1976] Robbins, S. E., "Time and Memory: The Basis for a Semantic-Directed Processor and its Meaning for Education." Doctoral thesis, University of Minnesota, 1976.

[Robbins2012] Robbins, S. E. "Meditation on a mousetrap: On consciousness and cognition, evolution and time," *Journal of Mind and Behavior* (2012): 33(1), 69–96.

[Robbins2018] Robbins, S. E., "Bergson's Holographic Theory—33—*Zen and the Art of Motorcycle Maintenance*," https://youtu.be/RaKC29-zWL0?t=1

[Robbins2022] Robbins, S. E., "Instinct as Form: The Challenge of Bergson," in Anne Malasse (Ed.), *Self-Organization as a New Paradigm in Evolutionary Biology: From Theory to Applied Cases in the Tree of Life,* New York: Springer, 2022.

[Roli2022] Roli, A., Jaeger, J., & Kauffman S. "How organisms come to know the world: Fundamental limits on artificial general intelligence," *Frontiers in Ecology and Evolution* (2022): 9, 1–14.

[Shermer2007] Shermer, M. *Why Darwin Matters: The Case Against Intelligent Design.* New York: Times Books, 2006.

[Vervaeke2019] Vervaeke, J. "Ep. 28. Awakening from the Meaning Crisis—Convergence to Relevance Realization," July 26, 2019, You-Tube video.https://www.youtube.com/watch?v=Yp6F80Nx0lc&/list=P LND1JCRq8Vuh3f0P5qjrSdb5eC1ZfZwWJ&index=29

[Vervaeke2023] Vervaeke, J. "John Vervaeke: Artificial Intelligence, the Meaning Crisis and the Future of Humanity, May, 2023, YouTube video. https://www.youtube.com/watch?v=VAiu7MR8Zqk

[Woodward2000] Woodward, J. "Explanation and invariance in the special sciences," *British Journal for the Philosophy of Science* (2000): 51(2), 197–214.

[Woodward2001] Woodward, J. "Law and explanation in biology: Invariance is the kind of stability that matters," *Philosophy of Science* (2001): 68(1), 1–20.

CHAPTER 6

REACHING FOR CUPS—
VOLUNTARY ACTION

LASHLEY AND THE SYNTAX OF AN ACT

The problem of voluntary action can be stated simply: how do we will to move our finger? Variants are: how do we will to swing baseball bats, or leap at mice, or reach out and grasp wing-flapping flies? Or speak a sentence? For a 4-D model of being, transforming with the rest of the universal field in an indivisible flow of time, this is a problem, for it is an uncharted path in a new, temporal, metaphysic of matter and mind. For the robotic framework, this is no problem— we simply turn on the power source of the robot-machine and let it begin executing its code instructions, and if the instruction is, "Grasp the coffee cup," it will move its mechanical arm, via a lot of programming, to grasp that coffee cup.

This is nice for the robotic framework and the machine theory of mind, but if this ease of accounting for action is taken as evidence for this theory as a veridical simulation of human action, it flies in the face of so many observations, as we shall see, that the ease of explanation becomes an obvious illusion.

But in the context of our previous discussion of perception and action, let us level set, firstly, for the seriousness of what faces AI in any claim that it is replicating human action. The principle of virtual action, with its intrinsic relativity, envisions a level of integration of perception and action beyond any theory today, clearly beyond the standard notion of the "perception-action cycle." No robot, as currently conceived, can implement this relationship. In the context of the Turing Test, therefore, expanding the nature and method of the test only slightly, we should be able to say to our robot, "Sir Robot, please take this little pill (containing various pertinent catalysts, but of strength undeclared to you) which will change your capability of action. Now please describe your perception." Given we have done previous calibrations on the strength of the catalyst regarding the perception-action effects on the average human, we would expect the robot, say, to begin speaking of heron-like flies, or, were the robot initially "perceiving" a cube rotating with sufficient velocity such that it is a fuzzy cylinder (a figure of infinite symmetry), he now should begin speaking of a rotating serrated-edged figure of 4n-fold symmetry (for he must continue to perceive by the same invariance laws specifying the event). Should we say, "Sir, please ingest a little more catalyst," he should begin speaking of a *cube* (a figure of 4-fold symmetry) in slow rotation, and should we ask him to ingest a bit more, he begins speaking of a stable, motionless cube or perhaps a stable, motionless fly "stuck" in a position over the coffee cup. And should we ask him to now reach for the cube, he modulates his grasping apparatus to grasp stable edges and sides, as opposed to a cylindrical figure, or for the fly, to slowly reach out and grasp a wing. That is, this is what we should expect of a being—and of the action of that being in different scales of time—for whom perception is virtual action.

We begin by turning again to the ubiquitous (at one time as far as subjects covered) Karl Lashley. Our bodily action involves the unfolding of ordered elements in time. These "elements" might be the words of a sentence or the various muscular movements underlying a dance step, or a tennis serve. The ordering or order of the elements is called the *syntax of the act*. Lashley pointed the field initially into the problem in his 1951 treatise, "The Problem of Serial Order in Behavior" [Lashley1951]. To pronounce the word "right," he noted, consists in these elements:

- The retraction and elevation of the tongue, expiration of air and activation of the vocal cords
- Depression of the tongue and jaw
- Elevation of the tongue to touch the dental ridge, stopping the vocalization and forceful expiration of air with depression of the tongue and jaw (see [Lashley1951], p. 187)

He noted that these movements have no intrinsic order of association. "Tire" requires these same movements in reverse order (or syntax). The order is imposed by some organization other than direct associative connections. This is equally true for the letters of a word. The letters "*R*" or "*G*" or "*A*" can occur in any order or combination. The order depends upon a *set* for a larger unit of action, in this case the word.

Words stand in the same relation to a sentence as letters to the word. Words themselves, he noted, have no *temporal valence*. The word "right" is noun, adjective, adverb, and verb, has four spellings and at least ten meanings. Take this sentence from Lashley:

> The millwright on my right thinks it right that some conventional rite should symbolize the right of every man to write as he pleases. ([Lashley1951], p. 187).

The arrangement of the individual elements or words, the order, is determined by a broad scheme of meaning. This holds not only for language, but for all skilled movements. What then determines the order, Lashley asked? Lashley took the view that the set or the idea does not have a temporal order; that all its elements are *contemporal*. Thus, we read a German sentence, pronouncing the words with no thought of their English equivalents, then proceed to give a free translation in English without remembering a single word of the German text. Lashley stated: "Somewhere between the reading and the free translation, the German sentence is condensed, the word order is reversed and expanded again into the different temporal order of English" ([Lashley1951], p. 189).

Lashley argued that the mechanism which determines the serial activation of the motor units is relatively independent both of motor units and of the thought structure. Supporting this, he pointed to a series of examples of mistakes of order. He had been tracking typing errors for some time. Misplacing or doubling a letter (t-h-s-e-s for these, i-i-l for ill, l-o-k-k for look), he argued, show the order dissociated from the idea. Further there are "contaminations," often from anticipation, for example typing "wrapid writing," or the Spoonerisms made famous by a probably somewhat aphasic Professor Spooner, for example, "Let us always remember that waste makes haste." These latter, he argued, indicate that a set of expressive elements seem to be partially readied or activated prior to the overt act.

Thus, Lashley argued (p. 192) that there are at least three sets of things to be accounted for:

1. The first is activation of the expressive elements (words or adaptive acts).

2. The second is determining tendency, set, or (atemporal) idea.

3. The third is the syntax of the act—a generalized pattern or schema of integration which may be imposed on a wide range of specific acts (upon individual components of an act).

So, bear in mind the "contemporal" nature of these elements in Lashley's account, and that atemporal idea. Let us see what became of this in Chomsky.

CHOMSKY AND SYNTAX OF THE SENTENCE

In the 1960s, the great linguistic theorist, Noam Chomsky, inspired by Lashley, would initiate a vast assault on point (3) in Lashley's "three sets of things to account for," the theory of syntax, overturning Skinner, his rats, and his simple "stimulus-response chain" Behaviorism as he did so and launching the cognitive science framework of psychology. He proposed a rule system or grammar, employed by the brain, to unfold the words from the idea in the proper order (Figure 6.1). The word-elements of the sentence, "The man stirred the coffee" are unfolded in a certain, grammatical order by a set of syntactic rules.

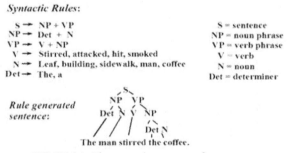

Syntactic Rules:

S → NP + VP
NP → Det + N
VP → V + NP
V → Stirred, attacked, hit, smoked
N → Leaf, building, sidewalk, man, coffee
Det → The, a

S = sentence
NP = noun phrase
VP = verb phrase
V = verb
N = noun
Det = determiner

Rule generated sentence:

The man stirred the coffee.

FIGURE 6.1. The syntactic structure of a sentence.

A sentence S, for example, by rule, may consist of a noun phrase and verb phrase ($S \Rightarrow NP + VP$), and a noun phrase consists of a determiner (Det) and a noun (N), and so on. The end-elements of the rule structure (nouns, verbs, determiners) come from the pool of possible words or symbols in the language. But there was always a profound implication in Lashley that Chomsky appeared to assume, and the subsequent seventy or so years of cognitive psychology has ignored.

At the apex of every diagram is "S." What is behind S? Behind S is, of course, the idea. It is the starting point from which every sentence derives its ultimate form, its syntactic structure, its embodiment in specific end-elements or words, that is, components of the sentence as an action. Behind the idea is a multi-modal event and its invariance structure. So, this event can be expressed in multiple degrees of complexity, richness, and detail. From the sparse, "The man stirred the coffee," the same idea could have been expressed as:

- The man quickly stirred the coffee.
- The arthritic old man quickly but painfully stirred the coffee.
- As the man stirred the coffee, the aroma wafted everywhere.
- The man quickly stirred the coffee with the spoon clinking away against the cup.
- The man stirred the coffee, sensing the metal surface of the spoon, the slight resistance of the coffee liquid to his push of the spoon, the nearly imperceptible scraping of the spoon against the ceramic cup surface, the surface swirl and the cream color slowly blending with the light brown of the coffee, with the slight heat felt coming into the spoon handle.

Cognitive science has pursued Chomsky's symbols-and-rules approach for over sixty years. We cannot imagine Lashley being impressed. To begin with, we have ignored his starting point, the atemporal idea. No post-Chomskian computer translation program, to include the GPTs, in all our subsequent syntactic sophistication, has ever translated a German sentence into an atemporal idea, then freely into English. No computer simulation program, in fact, has ever had an atemporal idea. We have had no theory of "S," the point from which syntactic structures must devolve in an ordered act.

In fact, cognitive science, as DORA for example (Chapter 4), has tried vainly to reduce "S" itself to data elements and yet more syntax. Yet it is from "S" that the expressive elements must be activated upon which this syntax operates. There is no theory of this activation given the nature of "S." "S"—the invariance structured event—sits in the realm of the commonsense knowledge problem—yet unreachable by AI. From this perspective, the attack on the true problem of syntax has not yet begun.

ATEMPORAL IDEAS AND THE DYNAMIC SCHEME

Dream literature is noted for viewing dreams as atemporal forms. Reports from personal experiences describe entire sequences of events taking place in what are apparently fractions of a second in normal time. Bergson describes a case from one individual:

> "I am dreaming of the Terror; I am present at scenes of massacre, I appear before the Revolution Tribunal, I see Robespierre, Marat, Fouquier-Tinville ... ; I defend myself, I am convicted, condemned to death, driven in the tumbril to the Place de la Revolution; I ascend the scaffold; the executioner lays me on the fatal plank, tilts it forward, the knife falls; I feel my head separate from my body, I wake in a state of intense anguish, and I feel on my neck the curtain pole which has suddenly got detached and fallen on my cervical vertebrae, just like the guillotine knife. It had all taken place in an instant, as my mother bore witness ... ."
> (Dreams, in *Mind Energy*, p. 129, [Bergson1912a], quoting Maury, *Le Sommeil et les Reves*)

The "man stirring coffee with the spoon" is inherently a time-extended event, however, as a thought, what is its duration? If our intent is to stir the coffee with the spoon, the elements may be conceived as atemporal, in fact an invariance structure defined over 4-D experience. Mozart once reported that in his creative process, he would see an entire piece of music, "though it be long," such that, "I can survey it, like a fine picture or a beautiful statue, at a glance." He adds, "Nor do I hear in my imagination the parts successively, but I hear them, as it were, all at once" [Mozart]. He was speaking of a time-extended, yet atemporal structure. Indeed, although we speak of 4-D extended events, the 4-D memory of which have been treating and its events have no particular duration. The detailed theory of the unfolding of an atemporal idea (as a physical action) is a challenge.

FIGURE 6.2. Learning to dance—the coordination of elementary movements via a dynamic scheme.

Bergson [Bergson1912b] treated some aspects of this in his treatise on *dynamic schemes*. Imagine learning to dance, he asked. To do so, we begin by watching people dance. The result is a visual impression of the movement. It is not a precise image, for the image has yet to be concretely filled in, articulated, so to speak, by the physically performed movements that will comprise the ultimate fluid action. These movements are elementary units that will be coordinated— movements used in walking, lifting up on one's toes, swinging the arms, and so on. But the first impression is an outline of the relations, especially temporal, but also spatial, of the successive parts of the movement. This *abstract image of the spatial-temporal relations* is the dynamic scheme.

This abstract scheme, he argued:

> ...must fill itself with all the motor sensations which correspond to the movement being carried out. This it can only do by evoking one by one the ideas of the sensations, or,

> in the words of Bastian, the "kinesthetic images," of the partial
> elementary movements composing the total movement:
> these memories of motor sensations, to the extent that they
> are revivified, are converted into actual motor sensations,
> and consequently into movements actually accomplished.
> ([Bergson1912b], pp. 217–218)

The preceding passage of Bergson contains, in a nutshell, the deepest
questions facing this model which we have been describing when it
comes to the frontier of voluntary action. Firstly, the scheme is not,
and cannot be, stored in the brain (for no *images* are stored there).
It is a creature of 4-D memory. Therefore, we are again coming
to grips with the nature of the whole transforming matter field of
which any individual being is a part.

Correlatively, it is not for nothing that Bergson's article is
entitled "Intellectual Effort." He is speaking of the *effort* needed to
transduce an abstract scheme into concrete images, images which
in turn are *inducing movements*. Effort is another term for *force*,
and this in turn sets us into the problem of just what force is once
we probe beyond Newton's definition. Finally, we have the picture
of the images integrally wound, causally, into the motor unfolding
of the act.

THE DYNAMIC SCHEME—LIBET'S EXPERIMENT

The lack of this conception in current theory, where the atemporal
idea unfolds into action and the larger vision of the nature of the
"subject" this implies, was the core of a significant controversy,
to this day very unresolved. In a famous 1985 study by Libet
[Libet1985], subjects were asked to perform an act like pushing a
button at their will. They were to note the point in time, by watching
a special oscilloscope display, when they had the conscious intent to
do the act. The studies reveal that there is a significant buildup of
neural activity 300 milliseconds before the time the subject reports
the intent to act.

Given that the intent is reported *after* the neural activity
buildup, the finding led to an ongoing controversy over the actual
role of free will, for it could be argued that the subject has no free

will at all and is simply reporting an "intent" that is nothing more than an unconscious, determined action already unrolling. It also led to Libet's suggestions—perceived as controversial—of "backward referral in time" of the intent.

But the gradual production of neural processes by an (initially "unconscious") idea, such that it eventually, via this means, becomes conscious, is an integral part of Bergson's model, for it is only by this means—the participation in a state of action/processing in the brain—that an idea can become conscious, and conscious as an image. From this perspective, other than showing that there is much we do not understand about the "unconscious," it is hard to see how Libet's result is so unexpected. Rather, it is exactly what should be expected in a larger, four-dimensional view of mind, where the "subject" is a time-extended being.

ACTUALIZING THE IMAGE

In one place, Bergson speculated to this effect, namely that matter (the brain architecture) itself might be delicately arranged, like a "trigger," allowing the image to affect it, causing the mechanisms underlying an act to unroll. This seems to introduce a point of confusion, of real dualism. It has images (mind) acting on brain (matter). A lapse in his thought? His indivisibly transforming holographic field is physical/psychical at once! We have a continuum: Matter (null scale, instantaneity, homogeneity, an ideal limit) ⇒ (ever-increasing scale of time) Mind. This framework we will examine shortly, but for now we stay simply in this framework of images driving action.

The notion of images driving an action was at one time quite normal in what was cognitive science before the advent of the computer and the (more) official birth of this science. William James (1890) and his ideo-motor theory was a dominant conception [James1890], at least until B.F. Skinner, his Behaviorism and his rats (wherein *anything* the brain did was expunged from study).[1] This

---

[1] In the author's experience at the University of Minnesota, cognitive psychology did not much exist even in 1967, only his final course given by a visiting professor from Michigan in 1967 even introduced the subject of cognitive development and Piaget. Several years later, in 1972, again at Minnesota, Newell and Simon's book served as the impetus for the all-out exploration of the computer metaphor of mind.

notion (images) is now equally anathema to cognitive science. This is precisely because the image has neither an explanation of its origin nor a useful place in cognitive science. As we saw, Pylyshyn, already in 1973 (only one year after Newell and Simon's 1972 *Human Problem Solving*) had reduced all mental phenomena to simply data elements and rules. For the neural network arm of cognitive science, there is only strength of connections between neural network units or nodes. Marc Jeannerod, however, reintroduced the concept of images driving actions in a 1994 article in the prestigious journal *Behavioral and Brain Science* [Jeannerod94]. Lacking a theory of the origin of images in the first place, Jeannerod was slightly ambiguous as to their actual role, for if experience is merely stored in the brain, images are simply generated from these stored elements. What role do they serve? As with Pylyshyn, they appear redundant (or "epiphenomenal").

Jeannerod took particular note of the research literature on the real effects of mental imagery practice on motor skill. These studies required the experimental subjects to practice a certain action in their imagination. After some number of these purely mental practice sessions, they were tested to see if they improved their speed and/or accuracy at the task. The studies show positive results. But there are studies in this context with more radical implications.

In the early 1950s, two researchers, Linn Cooper and Milton Erickson [Cooper52] [Cooper54], published the results of a series of experiments they had performed using hypnosis with imagery practice instructions (*Time Distortion in Hypnosis: An Experimental and Clinical Investigation*). The subjects were taught how to self-induce a trance to "alter time." In this altered-time trance state, they would be asked to simply imagine events or else to practice actions. For the one experiment they did on motor skill improvements in these practice sessions, greatly condensed in time, their results were marginal. Yet, curious incidents were observed.

One subject was mentally watching a movie in this altered time state. He was observed to be making extremely rapid movements with his hand, moving between his mouth and legs. When asked after the session what he was doing, he reported that he had "been eating popcorn." The velocity of his actions had been commensurate with the scale of time in his trance-altered state. Another subject, a violinist, was mentally practicing violin passages in altered time. She reported that in this state, she was able to view the whole piece, as though it were laid out as a static structure. We are reminded of Mozart's seeing an entire symphony as though it were a statue.

In our original discussion of altering the timescale and action within this scale (Chapter 3, "Bergson and the Image of the External World"), we relied on the physical effect of catalysts (i.e., from the *side of the object*—chemical changes, as in LSD, in the processes of the body/brain). In Cooper and Erickson's studies, we are seeing indications of similar effects, but from the "side of the subject" (i.e., via a purely "mental" impetus). There are other phenomena in everyday experience that open the same question.

Take for instance, the case of the great baseball player, Ted Williams (Figure 6.3). He was arguably the greatest hitter in baseball. He is the last hitter (to this date) to hit over .400 (.406 in 1941), a mark still not reached again for over eighty years now and counting—despite the interim advent of steroids. He also hit in

FIGURE 6.3. Ted Williams (1918–2002).

the .400's in 1952 and 1953. The author's father (a New Englander) loved to relate how Williams could read the label on a record as it spun around on the phonograph turntable. Williams could also, it was reported, see the seams and the spin on the baseball as it hurtled towards him. Williams, it was said, could "wait longer" than other players before he initiated his swing.

What does this sound like? It sounds, for all intents and purposes, as though Williams was in our "higher energy state." Watching a fly, he would have seen it moving more slowly—precisely because he could act more quickly. He "could wait longer" to swing because he was seeing, in the ball's slower motion, just how he could act, that is, virtual action, just like our cat watching the mouse.

A graduate school friend of the author, years ago, related that when he played basketball, on occasion, when driving to the basket, everything would slow down—the motions of the other players, the ball—and he would weave unimpeded to the basket. On other occasions, in baseball when batting, the ball would slow down and seem very large, and he seemed to have all the time in the world to smash a hit. Similar stories are not uncommon with people in the midst of accidents such as car crashes. The latter perhaps—the accident victims—can all be put down to adrenaline, but Ted Williams, the basketball playing grad student, the time-altered subjects, not so simply.

To cognitive science, in its limited, classic metaphysic-based concept of the brain, the question means nothing. But the studies are very real. How could this be possible? Of course, the "this" of this question is already: how can images drive the brain to action? We should note, however, that the "image" here in Bergson's framework is only a latter manifestation of the *intent* to act, only actualizing itself as an image as the various structures or areas of the brain required for the coordinated act swing into gear. This does not alleviate the "how can the mental be acting on physical?" form that this question-schema takes, for the "intent" nevertheless appears to be a *force*, a mental force, driving the brain. So, it will come down to these questions: what is force? Is *time* itself a *force*?

Let us already preview the "force" of the "Is time a force?" question. We have emphasized, particularly when discussing the thought of Roli et al. (Chapter 5) on the difference between organism and machine (along with Maturana and Varella, and also Mossio and Moreno), that something is being neglected, namely, that an organism is intrinsically embedded in the indivisible transformation of the universal field. This is to say that the organism is an integral participant in this transformation. In a metaphysical frame (the classic) wherein the universe is a field of independent, separate objects where "force" is required to impart "motion" to the objects, then the body and its internal structure down to its own tiny objects—the atoms—is just more of the same. In a universal field where the body—an organism—is an integral participant in this indivisible transformation, then the dynamics of action and the meaning of "intent" and "force" have a different basis. In this, too, we will be taking a deeper look into the difference between organism and machine than the previously mentioned authors have considered, for to them, *time* has not been the remotest of questions.

ORGANISMS: EXISTENCE AND CHANGE

In *Creative Evolution*, Bergson sets the framework of the discussion:

> ... for a conscious being, to exist is to change, to change is to mature, to mature is to go on creating oneself endlessly. Should the same be said of existence in general? ([Bergson1907], p. 7)

Take as an example of a "material object" a cube made of Lego blocks. Change it via an external force: we get a displacement of parts (Figure 6.4). If the parts change, these too are seen as composed of parts (and displaced), and so on, until we stop at the "unchangeable," that is, some agreed upon, hypothetical, minute particle. When a part leaves its position, nothing prevents its return, so a group of elements can always find its way back to the original state (by itself, or by some external force). Any *state* can be *repeated*. Implication: The group does not grow old. It has no *history*.

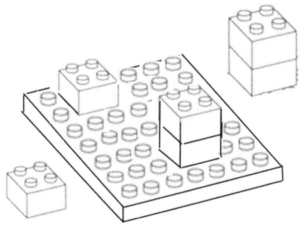

FIGURE 6.4. The Lego block "object."

Nothing, Bergson noted, is *created* in this. A superhuman intelligence (as per Laplace), computing the velocities of each part, can predict the future form. The formation and reformation take place in the abstract time of the classic metaphysic, a "time" which is only the noting of simultaneities in space. As we noted in the discussion of the cat's "internal clock" versus the physicist's standard measure of time, perhaps, using one revolution of a turntable as a "second," by convention each coincidence of a mark on a rotating disc with a mark on the frame on which it turns (Figure 6.5) is taken a "one second" — a pure spatial simultaneity, just as each coincidence of earth and sun at a certain point is taken as "one year."

FIGURE 6.5. The turntable. Each time a mark (arrow) on the rotating disk coincides with a mark (arrow) on the turntable frame, by convention, for example, one second has passed.

Speed up the turn table 2X: The time measure is unchanged. The same spatial simultaneities still occur. Abstract time does not care what happens in *the interval*! Only at the *ends* of the interval (i.e., at the mathematical point of a spatial simultaneity). Speed up the entire universe. Make the speed infinite, that is, so fast, it lays out like a "block" in space. The measured "time—the mathematical "time" —does not change.

FIGURE 6.6. The glass of water, an abstracted portion of the universal field.

But there *is* succession. As Bergson noted, place a cube of sugar in a glass of water; we must *wait* for the sugar to dissolve (Figure 6.6). This "waiting time" is *not* the mathematical time. It is *not* the time that would apply to the entire history of the material world. "It is no longer something thought, it is something lived. It is no longer a relation, it is an absolute" ([Bergson1907] p. 9). Thus, he notes:

> What else can this mean than that the glass of water, the sugar, and the process of the sugar's melting are abstractions, and that the Whole—within which they have been cut out by my senses and understanding—progresses, it may be in the manner of a consciousness? ([Bergson1907], p. 9.)

Note that this is a restatement from *Matter and Memory*, wherein we noted that the Whole is transforming, that "objects" and their "motions" are better seen as *changes or transferences of state* within this global motion. Thus, matter has a *tendency* to constitute isolable systems, again, as an *ideal limit*. These can be treated geometrically (as we will discuss), but the isolation is never complete:

> The thread attaching it [the glass, the sugar] to the rest of universe doubtless is very tenuous. Nevertheless, it is along that thread that is transmitted down to the smallest particle of the world in which we live the duration immanent in the whole of the universe. ([Bergson1907], p. 9)

And he notes: "The systems marked off by science *endure* only because they are bound up inseparably with the rest of the universe" (p. 9). Then we hit the *Matter and Memory* theme we have seen, that is, why the outlines of an object are only *practical* divisions, ultimately, as we shall see (Chapter 8, "The Problem of Affect") why *subject* must be treated as "different" from *object* in terms of *time*, not space, that is, not in terms of actual spatial division, and we see here the origin of the affordance (virtual action):

> … the distinct outlines we see in an object … are only the design of a certain kind of influence that we might exert at a certain point in space … the plan of our eventual actions … . Suppress this action … and the individuality of the body is reabsorbed in the universal interaction … ([Bergson1907], p. 9.)

ORGANISMS AS AGING

It is the very essence of *mechanism* (read the machine theory of mind here as well): "*... to consider as metaphorical every expression which attributes to time an effective action ...*" (p. 16). Concrete duration—the indivisible, concrete transformation of the field, where the (present) instant is the reflection of (and integrally bound with) the entire preceding series—this is denied. Back then to our "Lego" objects:

> Change must be reducible to an arrangement and rearrangement of parts; the irreversibility of time must be an appearance relative to our ignorance ... the impossibility of turning back must be only the inability of man to put things in place again. So growing old can be nothing more than the gradual gain or loss of certain substances ... "
> ([Bergson1907], p. 16)

Time has no more reality than an hourglass: just flip it to return things back to the start. The cause of aging must lie deeper:

> The impetus which causes a living being to grow larger, to develop and age, is the same that has caused it to pass through the phases of embryonic life ... It is evident that a change like that of puberty is in course of preparation at every instant from birth ... (p. 18)

For an unorganized body (i.e., an inorganic body), its present state (or present instant) depends entirely on the state of the previous instant. As an isolated system, the position of its points is determined by the position of the same point at the previous instant, thus, by equations with coefficients: ds/dt, dv/dt, and so forth. Bergson asks, is this so with the laws of life? "... does the state of a living (organic) body find its complete explanation in the state immediately before?" (p. 19).

Yes, if there is an a priori agreement to:

- liken the living body to an inorganic body
- treat the living as the systems studied by the physicist, chemist, or astronomer

But for the organic, this mathematical treatment is not possible. No, this is not due to our ignorance, but rather, for a living body:

- The present instant is not explained by the immediate previous instant.
- ALL the past of the organism must be added to that moment.

This is the cardinal difference between *abstract* time and *concrete* time:

> ... the idea that a living body might be treated by some superhuman calculator *in the same mathematical way as our solar system* ... has gradually arisen from a metaphysic [the classic] which has taken a more precise form since the physical discoveries of Galileo ... *its apparent clearness, our impatient desire to find it true*, the enthusiasm with which so many excellent minds accept without proof ... should put us on our guard against it. (p. 22, emphasis added)

The instant "immediately" before is connected with the present instant by the interval, dt. Thus, the present state is defined by equations with differential coefficients: ds/dt, dv/dt, or in other words:

- by present velocities and present accelerations
- a present considered along with its tendency

The physicist is concerned with the instant/point of the coincidence of the two "marks," one on the rotating disk and one on the platform of the turntable of Figure 6.5, not the interval involving the rotation of the disk/mark or point as it travels around. It is the t-th moment that counts, that is, the instant, not the interval, or rather the *extremity* of the interval—the instantaneous cross-section of a flow. The physicist can divide the interval, creating/defining infinitely small parts, that is, ever smaller intervals—(dt)s—and now computing accelerations, velocities, that is, *tendencies*. But this is still defining only static moments, not flowing time, again, a series of extremities or ends-of-intervals—"instants"—instantaneous cross-sections.

We are back to our figural representation of the classic metaphysic—the "Cubes" of the all of Space (Figure 6.7), each

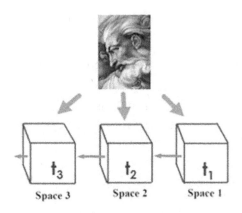

**Instantaneous Cubes
of the ALL of Space**

FIGURE 6.7. The instantaneous "Cubes" of Space, each corresponding
to a mathematical instant of time.

correspondent to a mathematical point/instant of time-extent. The
systems that science works with, notes Bergson: "... are in fact *in
an instantaneous present that is always being renewed* ... never in
that real, concrete duration in which past remains bound up with
the present" (p. 22). And as was earlier noted, such a Cube with
a mathematical instant/present of time-extent would be frozen,
no transition from Cube 1 to Cube 2 being possible without some
"external intervention," for example, an act of *creation* by a God.

> The world of the mathematician ... a world that dies and
> is reborn at every instant—the world which Descartes was
> thinking of when he spoke of continued creation. (p. 22)

Physics fools itself, he noted: it cannot escape *continuous creation.*
This instantaneity is an *abstract limit.* And the isolation (say, of the
water glass and sugar) is an abstract limit. Neither is reached in the
concretely transforming concrete world. No system is isolated from
the indivisible, global transformation.

THE LIMITS OF MATHEMATICAL LAW

But this assumption rests at the base of measurement, and thus, of
the laws of science. Bergson uses "boiling water on a stove" (Figure
6.8). Yes, the system can be *practically* isolated. But in reality: it is

bound up with other objects and operations. "In in the end, I should find that our entire solar system is concerned with what is being done at this particular point of space" (p. 214). To say, "The water will boil today, just as it did yesterday," I am vaguely imagining placing:

- stove (yesterday) on stove (today)
- kettle on kettle
- water on water
- duration on duration

Yesterday **Today**

FIGURE 6.8. Superimposing two boiling events.

But, to imagine this, to superimpose yesterday on today: yesterday's system had to *wait* for today's, that is, *time had to halt*, and everything became simultaneous. This happens in geometry, and only in geometry: we superimpose two triangles and show the two are equal (a purely spatial operation). In this inductive process, for both physicist and geometrician, time does not count. Bergson insists, "there is something artificial" in the mathematical form of physical law. "*Nature did not dream this superposition*" (p. 218). The mathematical order is nothing positive. "It is the form towards which a certain interruption tends of itself, and that materiality is precisely an interruption of this kind" ([Bergson1907], p 218).

We have hit a core within Bergson's vision. The vision is this:

- It is Consciousness—the intensive—that "pushes" or impels the universe into existence.
- This push is an intense, intensive force.
- The relaxation (lessening) of this force is the emergence of the extensive—the material world—with its geometric tendencies.

Thus, he states:

> ... whereas the power of creation once given (and it exists, we are conscious of it in ourselves ...) has only to be diverted from itself, to relax its tension, only to relax its tension to extend, only to extend for the mathematical order of the elements so distinguished and the inflexible determinism connecting them to manifest the interruption of the creative act ... (p. 217)

To put this differently, the abstract space of the classic metaphysic—the continuum of points/positions (Figure 6.9) with its abstract "time" of instants—is again the ideal limit—never truly reached—of this continuous "thrust" of the intensive order into extension, that is, into and as the concrete, physical world. Physics succeeds since, for although there is no definite system of mathematical laws, matter inclines to this (abstract) form, toward this limit. But as we go back up the course of the inner thrust, so to speak, to biology, then to psychology, this abstraction becomes increasingly less applicable to each science. The vision of Kant (Figure 6.10) wherein all sciences proceed on the same basis, that is, on the same abstraction presented by the classic metaphysic is not, and cannot be, valid. This is also to say, physics cannot be the sufficient basis from which starts the science of psychology and especially the problem of consciousness.

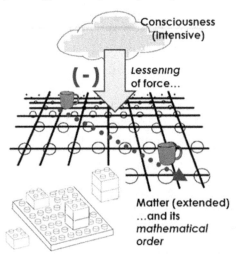

FIGURE 6.9. The extension of the abstract space, the infinitely divisible continuum of points/positions, as an *ideal limit*, a *lessening* of the force of Consciousness expressing as the material world.

We are going to examine this "intensive-to-extensive" conception only a little further, but it is important to have a conception of what this alternative vision is and what is operating behind the points being made here.

FIGURE 6.10. The Kantian unified structure of the sciences, all resting on the same (mathematical) basis of explanation.

Continuing with the nature of mathematical (physical) law, Bergson noted the transition from the ancient conception of law to that of the moderns. The ancients were focused on genera. For example, take Aristotle's law of falling bodies. This involved things like "proper place," "high, low," "forced movement," "natural movement." The stone regains its "natural place." In other words, it is like a living being—the genus "stone." For we moderns, laws are no longer reduced to genera, rather, genera are reduced to laws.

This goes back to the laws of Kepler and Galileo—a planet sweeping out equal areas over equal times, or the law describing the descent of a ball, instant by instant, down an inclined plane. But a law is *a relation* between things; it is expressing a magnitude as a function of several other (appropriately chosen) variables. It is a *comparison*:

- An "objective reality" only for an intelligence that represents to itself several terms at the same time
- That is, a relation between things
- Expressing a magnitude as a function of several other (appropriately chosen) variables

We shall have reason to view this again—the *anthropomorphic* nature of these laws—in Chapter 8. Kant, Bergson notes, only brought this to light. It is implicit in a universe of laws. So, we come to the 1st and 2nd laws:

▪ The 1st—Conservation of Energy—is quantitative, hence relative.

- It is tailored to measuring a certain form of energy

- It is concerned with a fragment of this world relating to another fragment—not to the nature of the whole. (Again, a comparison.)

▪ The 2nd—Entropy—is the most metaphysical of physical laws.

- The essence: All physical changes tend to degrade into heat.

- The heat is distributed.

- There is no artifice of measurement here.

The 2nd law tells us: "changes that are visible and heterogeneous will be more and more diluted into changes that are invisible and homogeneous" (p. 243). Dynamic change characterizes organisms or even our galaxy and solar system, and compared to a final, entropic state, they are the essence of instability:

> The instability to which we owe the richness and variety of changes taking place in our solar system will gradually give way to the relative stability of elementary vibrations continually and perpetually repeated. (p. 243)

From what place or source does this mutability come? This is still, of course, the question. Sean Carroll explored it [Carroll2009], opening his book with the question: "Why the apparent very low entropy at the start of the universe?" (i.e., before the Big Bang). We ask these kinds of questions: Does it come from some other point in space? This starts an infinite regression. An infinite universe? This admits the coincidence of matter with abstract space. And how would this latter be reconciled with the holographic vision wherein we have the reciprocal action of all parts of matter upon each other? Alternating phases of stability and instability? Boltzmann showed it was mathematically improbable—to the point of impossibility. The problem is insoluble staying upon the grounds of physics. The physicist must remain in space (the abstract space), noted Bergson. He belies his role if he seeks the source in an extra-spatial process. But we are dealing with *consciousness*; this is where we must go.

TIME IS A *FORCE*

It is not extension in the abstract we are trying to explain. Extension is a tension (intensive) interrupted. Again, it (the abstract space) is an ideal limit—never reached. We are explaining the *concrete* reality. There is an order that reigns there, it is an intensive order. We need only to think of Beethoven and the 9th Symphony—certainly an intensive order, a vital order—within the mind of a Beethoven, expressed (in *extended* form) only as it unfolded via the musical notation on his manuscript. So, the extensive order is born when the inverse order (the vital) is suppressed. This suggests, says Bergson, *a thing unmaking itself.*

> What conclusion ... if not that the process by which a thing makes itself is directed in a contrary way to that of physical processes, and that it is therefore, by its very definition, immaterial? ([Bergson1907], p. 245)

Take this again to that glass and the dissolving sugar. Why must we *wait*? We are conscious beings; we are not interested in the extremities of intervals. We feel and live through and within the intervals themselves. We are one with, participating in, the duration unfolding in the glass, the water, the sugar. That is, we are integrally participating in the creative unfolding of the universal field. Time— duration—is a force. The creative force.

This is why everything cannot be given at once, or at some incredible, arbitrary speed, like the projection of frame after frame via a movie projector, with near infinite velocity. Bergson thus states: "The more I consider this point ... if the future is bound to succeed the present ... it is because the future is not altogether determined at the present moment" (p. 339).

> It is because there is unceasingly being created in it, not indeed in any such artificially isolated system as a glass of sugared water, but in the concrete whole of which every such system forms a part, something unforeseeable and new ... (p. 339)

Finally:

> The duration [the paced unfolding] of the universe must therefore be one with the latitude of creation which can find place in it. (p. 339)[2]

MECHANICAL CAUSALITY AND REPEATABILITY

We can return now to the subject of *force* and the problem of the "cause" of voluntary action. The essence of the concept of *causality* in the classical framework is *repeatability*. Causality, as an explanatory concept in science, is useless if the cause is not repeatable. If we line up a cue ball in the same precise position relative to the eight ball, apply precisely the same vector of force via the cue stick, we expect the same result. Our prediction scheme via Newtonian laws works wonderfully. As Bergson argued in *Time and Free Will*, if we are faced with a choice (Figure 6.11), say, between a glass of Burgundy versus a glass of Bordeaux, the choice is in effect seen as a *trajectory*—an infinitely divisible line of instants/points that we follow, reaching an oscillation phase at point D, just like a pinball machine, then the path taken to Bx. All the causes acting upon both the physical and biological level are taken as exactly the same, thus we would envision exactly the same result—we invariably travel down the path from D to Bx and pick up the Bordeaux—and we say all is "determined."

In truth, all these things are only *practically* the same. The billiard ball is not the same from strike to strike. The cue ball is not

---

[2] McGilchrist [McGil2021], in his reflections on evolution, certainly is aligned with Bergson and a creative force driving the evolution of the universe. In this book, we have not attempted to integrate McGilchrist's massively documented findings and discussion of the different functions of the left and right hemispheres of the brain, although this is certainly another dimension of challenge for AI. We might simply note, though, the alignment of this hemispheric function-split with Bergson's "dichotomies:" Left brain: Abstract space, abstract time, quantity, syntax, intellect. Right brain: Extensity, duration, quality, semantics, intuition.

McGilchrist is clearly a Bergsonian, particularly in the conception of time and the nature of evolution, noting [McGilchrist2023] that Bergson "… is enormously underappreciated these days," and, "Yes, if I were asked which philosopher in the last 100 years it is most important to read, it would be Bergson."

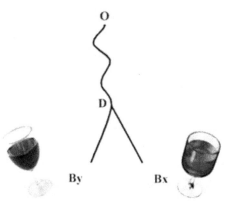

FIGURE 6.11. The choice between the Burgundy and the Bordeaux.

the same. The piano key, struck by our finger ten times in a row, is not the same piano key from strike to strike, nor is it the same finger. Nor the same pinball. Nor the pot with its boiling water. All are only *practically* the same.

But from whence comes the concept of repeatability? It only has meaning within the abstract space of the classic metaphysic. We saw that the abstract "Cubes" of Space have a duration so minute, so infinitesimally tiny in time, that our "Cube" of Space has nothing left of the qualitative aspect of the perceived world. In fact (as noted in Chapter 3), we are stripping all quality as we descend to increasingly smaller scales of time. The buzzing fly of our normal scale of time becomes the immobile fly, becomes a cloud of electrons, becomes a haze of quark events, becomes an ensemble of strings, becomes ultimately so many algebraical relations. We end with a virtually homogeneous, featureless, quality-less Cube, then another, then another ... We are close to achieving absolute repeatability, for we are absolutely near the concept of *quantity*.

When we count, for example, a set of apples, we strip each apple of any individuality, any differentiating quality; we ignore their individual differences; they are for this purpose homogeneous. We are only interested in the repeatability of the counting operation— apple 1 + apple 2 + apple 3 ... We have achieved our repeatability, but at enormous expense. How do we get Cube 1 to generate or "cause" Cube 2? How does Cube 1 "force" the existence of Cube 2? How, in other words, do we bind the future to the present? The truth is, just as both Lynds [Lynds2003] and Bergson (long before

Lynds) argued, we cannot. If the universe can be momentarily static, then it is a universe forever incapable of change.

The universe of Laplace and his demon rests upon this classical causality, which in turn rests on repeatability. We can only predict the positions and velocities of each particle at the next instant if each preceding instant is utterly homogenous and repeatable. But we can only attain absolute homogeneity by robbing the Cubes of all motion, therefore all extent in time. We end, at limit, again, with a universe that is incapable of change.

This leads to the question of force itself.

DYNAMICAL CAUSALITY, CONSCIOUSNESS, AND "FORCE"

The preceding—the "Cubes" —is the *static* framework of binding the past to the present (basically, *concatenating* the Cubes). But as Bergson noted [Bergson1889], there is yet the *dynamic* conception of the binding of present to past. It is derived directly from our conscious experience. We reach for the coffee cup. We have an idea of an action, we initiate the action, there is throughout the act the accompanying feeling of effort or force, and the action is completed. The whole sequence, from idea to act, is experienced as a continuous whole, and by this very wholeness, the act feels prefigured in the idea. Yet it is also our deepest experience that this is not a necessary connection—the act can be aborted. It is the source of our notion that psychological causes are "different" from physical causes. This dynamic picture of causality can in fact easily extend to the material universal field flowing in time, similar to our conscious experience, and where the future is not bound by necessity to the past.

But this indubitable experience of our consciousness, with its experience of force or effort, is projected upon the external world in a mixture with the concept of a Laplace-like necessary or mathematical determination. The dynamic and the static become subtly, unconsciously joined. In the union, Bergson argued, the concept of force is joined with necessity or with a necessary unrolling by mathematical law—*force now determines effects* [Bergson1889]. We then turn this upon ourselves, viewing our own consciousness through the very layer of forms we have projected, that is, via the projected layer of concepts of the spatial metaphysic. The mechanical, infinitely repeatable action of one object upon another, for example, cue ball upon eight ball, assumes the same form as the dynamic expression of force and resultant action experienced in consciousness when reaching for the coffee cup, while in turn this phenomenal experience itself is painted with an algebraic prefiguring and mechanical unrolling via the force that causes it. In other words, we have confused our experience of forcefully acting, which is integrally derived from our embeddedness in the indivisibly transforming universal field, with the concepts of the classic metaphysic.

In the classical metaphysic, we have seen that all motion is relative. An object can move from point to point across the continuum of positions, or the continuum (coordinate system) can move beneath the object. But we have also seen that there must be *real motion* in the matter field—stars explode, trees grow, and humans wrinkle and age. How can we distinguish real motion from that which is merely relative, and which rests upon a change of perspective? "Force" appears as the natural answer. Force is naturally seen, in our classic framework, as that which imparts "motion" to "objects." Real motion emanates from a "force." The "image" (better, *intent*) acting on the brain, would have to be a force—it would be the cue stick striking the cue ball.

But this is the problem: force is only a function of mass and velocity, where velocity is the rate of change of position. Force, or $f = ma$, is measured by the degree of acceleration it produces in the body (or mass, m). In turn, acceleration is only the rate of change of the rate of change of position. (Velocity is the first derivative of change of position with respect to time; acceleration is but the second derivative.) We are always dealing, in other words, only with *change of position*. In other words, these movements are still *relative*. The force, one with these relative movements, does not escape this relativity. Force is no more absolute than the movements; it cannot serve to distinguish real or absolute motion.[3] It is not a "cause." It is,

---

[3] Obviously, Einstein made precisely this move—assigning force, an *absolute force*—to acceleration. Bergson [Bergson1922] noted this problem in his comments on General Relativity within his examination of Special Relativity in *Duration and Simultaneity*, 1922.

expressed in $f = ma$, as physics is cautious to treat it, an invariance law. Thus, as Bergson, noted:

> It is in vain, then, that we seek to found the reality of motion on a cause which is distinct from it: *analysis always brings us back to motion itself.* ([Bergson1896], p. 257, emphasis added)

With the same caution, psychology would say that our *inner causality* has no relation to the mechanical effect of one object upon another. This latter can be conceived as capable of repeating in a homogeneous space, and thus expressible by law—cue ball repeats hitting eight ball, eight ball goes into the table pocket. But the organism and its consciousness flow in real time, its "states" can never recur; they are not truly repeatable, each "state" being a reflection of the preceding series. Further, as already noted, it must be realized that this is true for the motion of the entire matter field, the field in which our body and individual consciousness integrally participate. Only at a sufficient level of scale can we treat this field and its events as *practically* repeatable and subject to classical laws.

We come, then, to these implications: the very concept of physical force is derivative, a projected aspect of the physical/psychical evolution over time of the mind/matter field (via the cognitive "projection frame" of the classic metaphysic—itself a *derivative* of our perception's identification of separate "bodies" on which we can act, and then on subsequent cognitive development). Our action, from intent to concretely grasping the spoon and stirring, intrinsically participates in this psychical-physical flow. Certainly, there is much more to understand here, but this is the framework we require for a theory of voluntary action. Language—speaking our thoughts—is one very clear manifestation of voluntary action, and the ideas expressed. Even the spoken words expressed with their own multiple coordinated articulatory movements, just as in a dance, are dynamic schemes. It is to language that we now turn.

REFERENCES

[Bergson1889] Bergson, H. *Time and Free Will: An Essay on the Immediate Data of Consciousness.* London: George Allen and Unwin Ltd., 1889.

[Bergson1896] Bergson, H. *Matter and Memory.* New York: Macmillan, 1896/1912.

[Bergson1907] Bergson, H. *Creative Evolution.* New York: Holt, 1907/1911.

[Bergson1912a] Bergson, H. "Dreams," in *Mind-Energy, Lectures and Essays.* (W. Carr, Trans.). New York: Holt, 1912/1920.

[Bergson1912b] Bergson, H. "Intellectual effort," in *Mind-Energy, Lectures and Essays*, (W. Carr, Trans.). New York: Holt, 1912/1920.

[Bergson1922] Bergson, H. *Duration and Simultaneity With Respect To Einstein's Theory.* Indianapolis: Bobbs-Merrill, 1922/1965.

[Carroll2009] Carroll, S. *From Eternity to Here: The Quest for the Ultimate Theory of Time.* New York: Dutton, 2009.

[Cooper52] Cooper, L. F., & Tuthill, L. "Time distortion in hypnosis and motor learning," *Journal of Psychology* (1952): *34*, 67–76.

[Cooper54] Cooper, L. F., & Erickson, M. *Time Distortion in Hypnosis: An Experimental and Clinical Investigation.* Baltimore: Williams & Wilkins, 1954.

[James1890] James, W. *Principles of Psychology.* New York: Holt. 1890.

[Jeannerod94] Jeannerod, M. "The representing brain: Neural correlates of motor intention and imagery," *Behavioral and Brain Sciences* (1994): *17*(2), 187–245.

[Lashley1951] Lashley, K. "The problem of serial order in behavior," in A. Blumenthal (Ed.), *Language and Psychology* (183–193). New York: John Wiley & Sons. 1951.

[Libet1985] Libet, B. "Unconscious cerebral initiative and the role of conscious will in the initiation of action," *Behavioral and Brain Sciences* (1985): *8*(4), 529–566.

[Lynds2003] Lynds, P. "Time and classical and quantum mechanics: Indeterminacy versus discontinuity," *Foundations of Physics Letters* (2003); *16*, 343–355.

[McGilchrist2021] McGilchrist, I. *The Matter With Things: Our Brains, Our Delusions and the Unmaking of the World*. London: Perspectiva Press, 2021.

[McGilchrist2023] McGilchrist, I. "Understanding the Matter With Things—Dialogues—Episode 22: Chapter 22, Time," February 27, 2023, YouTube video. https://www.youtube.com/watch?v=2KRs664HLb0

[Mozart] https://libquotes.com/wolfgang-amadeus-mozart/quote/lbs0e8s

CHAPTER 7

GENERATIVE *AI* AND HUMAN SPEECH

PRELIMINARY: BERGSON AND SPEECH

We are going to look at Bergson's model of language comprehension, placed within the context of AI speech perception and the generative pretrained transformers (GPTs). To set the stage, we note that there is no current, actual understanding of human language understanding, just as there is none for perception, thus for memory. So, language, dependent on both, is going to be no better. Machine/AI efforts, like SIRI, are not even attempting to replicate humans. So, in this subject, there is a massive explanatory "gap."

Where does Bergson's model fit in his overall pursuit? Recall his main question in *Matter and Memory*: *Is experience stored in the brain?* The answer, we noted, is determinative of a theory of consciousness. As noted earlier (Chapter 2), virtually no one today (Sheldrake as one, partial exception) understands the significance of this question. No one poses it. No one considers it, not Searle, not Chalmers, not Dennett, not Block. With no memory model for the simplest of events—just stirring that coffee—there is no theory of consciousness!

So, Bergson was researching memory "losses" —aphasias, amnesias. One class of these is loss of the ability to understand language. Sometimes there is a loss of individual words, sometimes classes of words. Does this imply "words" are stored in the brain? To answer this, he had to build a model of language understanding, not in great detail, but sufficient to address the "storage" question.

So, the model is not detailed, but takes its place within his overall holographic framework, thus a challenge to the current mindset and the gap.

The plan here is to see just where we are in terms of our understanding, first of speech perception, and second, its inextricable correlate, language understanding, especially when taken in the context of the vast efforts in the last half-century plus in machine perception and understanding of language, to include the GPTs or generative AI.

MACHINE SPEECH RECOGNITION (1950S TO THE GPTS)

The speech perception problem is this: how do listeners extract the most fundamental linguistic elements—consonants and vowels, or the distinctive features which compose them—from the acoustic signal? We are presented, so to speak, with an acoustic mass of frequency and amplitude fluctuation (Figure 7.1). How is this mass of sound segmented into the clear "words" of speech (where the words consist of phonemes—consonants and vowels)? Stated in this way, there is an implicit, limited objective in this problem, namely, we are just interested in taking the acoustic stream and outputting the correct sequence of words within the stream, for example, if the human speaker has said, "The man stirred the coffee," the output is indeed: [The, man, stirred, the, coffee] and in that order. We are not yet worried about what this string of words (as a sentence) *means*. Meaning comes (mostly) later.

The man   stirred   the   coffee.

FIGURE 7.1. The speech segmentation problem

This limited objective, it should be noted already, already implies a preconceived view of the problem. It implies a partition of the process into a set of *temporal stages* that may not exist: [process the phonemes, then the words, then the sentence, then meaning]. One can take this statement as a preview of the nature of the problem that has beset the field.

As noted, no attempt is being made to segment this stream in the way or process by which humans do this. We shall understand why shortly. In the 1950s, computer efforts began already at Bell Labs with a system for single-speaker digit recognition. It was limited (in the 1950s) to single-speaker systems and around ten words. It worked by locating the formants in the power spectrum of each utterance. Formants are the spectral peaks of the sound spectrum, the acoustic resonance of the human vocal tract. So, it is a resonance or the spectral maximum that the resonance produces, measured as amplitude peaks in the frequency spectrum of the sound, using a spectrogram.

In the 1970s, James and Janet Baker introduced the hidden Markov model (HMM). This combines different sources of knowledge—acoustics, language, syntax—in a probabilistic model. Already, at this point, we see the veering toward statistical modeling, away from emulating the human brain. Linguists at the time were not happy, feeling that the hidden Markov models are too simple to capture language. But the linguists were ignored, and in the 1980s, HMMs were dominant.

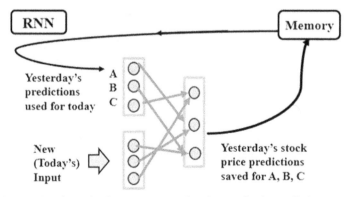

FIGURE 7.2. Schematic of a recurrent neural net in a stock price prediction context.

With the advent of deep learning in the early 2000s, the long short-term memory (LSTM) flavor of recurrent nets (RNNs) took over (Figure 7.2). The LSTMs handle cases requiring memories of events that happened thousands of discrete time steps ago (important for speech). The critical change that occurred in the 2000s of course was the advent of big training data and big computing power. This greatly increased accuracy over the HMMs, at least by 30%+ in the initial stages of this big data revolution.

The emergence of the GPTs and the transformer (encoder-decoder) architecture (cf. [Vaswani2017]), which does away with RNNs and LSTMs, has continued the trend. Radford et al. [Radford2022] describe the use of this architecture in their model, "whisper" (which, they say, is close to an acronym for "Web-scale supervised pretraining for speech recognition"). This model employs a large amount of "weakly supervised" data, this being comprised of 680,000 hours of labeled audio data, though the audio-to-transcription mapping is not up to the "gold standard" datasets of this sort. The "gold standards" have far more laborious human verification of their correctness and much reduced noise, but comprise far, far fewer hours of available data (roughly only 40,000 hours). In other words, like mapping the input digits 0–9 (i.e., the pixel patterns) to the correct neural net classification output units, the exercise is still focused on a *mapping* task—achieving the correct mapping of the written text to the acoustical stream, where the sound stream itself is initially digitized.

A human, listening to 680,000 hours of speech at eight hours per day and reading the transcripts, would require 232 years of training. Realizing the absurdity of making a claim that these algorithms are equivalent to what people do, AI is working on using smaller datasets, with improved, although roughly the same, class of algorithms. The bottom line here is that the AI community has had little interest in how people do it, and still is on the path of achieving something that will only vaguely appear to be, or will pass as, the human case. But just what were/are the main theories about how people do speech perception (about which the AI-folks are basically unconcerned)?

HUMAN SPEECH PERCEPTION: THE THREE THEORIES

There are three main theories on how we humans perceive that speech stream, parsing it into vowels and consonants, ultimately words and sentences:

- ▪ the motor theory of speech perception
- ▪ the direct realist theory (aligned with Gibson)
- ▪ the general learning approach

We begin with the motor theory.

The Motor Theory of Speech Perception

The motor theory (henceforth MT) started in the 1950s at Haskins Labs with Alvin Lieberman [Lieberman1967], [Lieberman1983]. It involved a series of landmark studies on the perception of synthetic speech sounds. As Figure 7.1 gives a rough idea, the mapping between speech signals (acoustics) and linguistic units (phonemes) is very complex. The conclusion of the Haskins studies was this: perceived phonemes and features have a simpler relationship to *articulation* (the vocal tract movements) than to the acoustic signal (i.e., more nearly one-to-one).

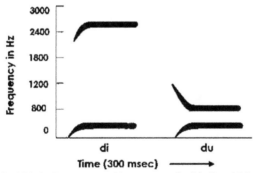

FIGURE 7.3. Formant transition patterns for "dee" and "doo."

Every version of the motor theory has claimed that the objects of speech perception are articulatory events rather than acoustic or auditory events. More specifically, it was hypothesized that the articulatory events recovered by human listeners are neuromotor commands to the articulators (e.g., to the tongue, lips, and vocal folds). This is rather than more peripheral events such as actual articulatory movements or (articulatory) gestures (where a "gesture" like waving one's hand in this framework is equivalent to a "gesture" such as moving one's lips). The objects of speech perception must be more-or-less invariant with respect to phonemes or feature sets and by a further belief that such a requirement was satisfied only by neuromotor commands.

Consider two synthetic formant patterns perceived by listeners as the syllables /di/ ("dee") and /du/ ("doo") in Figure 7.3. The steady-state formants (horizontal line) correspond to the target frequency values of the vowels /i/ and /u/. The rapidly changing (rising) formant frequencies (formant transitions) at the onset of each syllable carry important information about the initial consonant:

- The rising first-formant (F1) transition of both syllables signals that the consonant is a voiced "stop" such as /b/, /d/, or /g/.
- The rising second-formant (F2) transition of /di/...
- The falling F2 transition of /du/ provide critical information about place of articulation (i.e., that the consonant is /d/ rather than /b/ or /g/).

How does so different a pattern of F2 transition give rise to the same phonemic percept? This suggested that invariance must be sought at an articulatory rather than an acoustic level of description.

The Direct Realist Theory

The direct realist theory (henceforth, DRT) began in the 1980s with Carol Fowler [Fowler1986], also at Haskins. DRT asserts that the articulatory objects of perception are actual, phonetically structured vocal tract movements or gestures. It does not assign any effect to events that are causally antecedent to these movements (e.g., neuromotor commands). DRT also contrasts sharply with MT in denying that specialized (i.e., speech-specific or human-specific) mechanisms play a role in speech perception. Following Gibson, Fowler [Fowler1986] argues that speech perception can be broadly characterized in the same terms as, for example, visual perception of surface layout (as in the several figures of Chapter 2, e.g., Figures 2.4 and 2.5). She notes that animals use structure in the media that has been lawfully caused by events in the environment as information, but though it is the structure in media (light for vision, skin for touch, air for hearing) that sense organs transduce (e.g., the texture gradient structure with its invariance laws), it is not the structure in those media that animals perceive, they perceive the components of their niche (e.g., a nice, flat, walk-on-able surface) that caused the structure.

The term "direct" in direct realism is meant to imply that perception is not mediated by processes of inference or hypothesis testing. Rather, the information in the acoustic signal is assumed to be rich enough to specify (i.e., determine uniquely) the gestures that structure the signal. A talker's gestures structure the acoustic signal (e.g., the closing and opening of the lips during the production of /pa/) which then serves as the informational medium for the listener to recover the gestures. To perceive the gestures, it is sufficient for the listener simply to detect the relevant information. That /di/ and /du/ are perceived as having the same initial consonant (despite their disparate F2 transitions): both MT and DRT see this in terms of an assumed commonality of gestures in the two cases.

By this point it is becoming clearer why the AI-world does not care how humans do it. For either MT or DRT, it would seem that a robot must have the human vocal apparatus or something somewhat close (or the detailed *physical* understanding thereof).

The General Learning Approach

The general learning theory (henceforth, GLT), proposed by Randy Diehl [Diehl2004], is almost more of reaction to findings appearing to contradict MT and/or DRT than a well specified theory. Diehl's own conclusion:

> Therefore, a major challenge for GA is to develop hypotheses based on far more accurate information about the auditory representation of speech and the statistical properties of natural speech. ([Diehl2004], p. 173)

GLT does not invoke special mechanisms or modules (as does MT) to explain speech perception. It assumes speech sounds are perceived using the same mechanisms of audition and perceptual learning that have evolved in humans to handle other classes of environmental sounds. Speech perception is neither equivalent to nor mediated by the perception of [vocal] gestures.

GLT notes several problems that appear to afflict either MT, DRT, or both. For example, birds have been trained to discriminate /di/ versus /du/. The apparent implication: having the human vocal apparatus is not important (though, on the other hand, the finding hardly implies birds are doing the same thing as humans). There are several other phenomena, to include "categorical" perception and several phonetic context effects, where the argument is that these are not unique to speech or to human listeners.

Can the sound stream *precisely specify* its source (as per DRT)? First, consider the optical case: For organisms with two or three types of cone photopigments, the spectral distribution of reflected light cannot be unambiguously recovered. This is because every pattern of cone responses is compatible with an infinite set of hypothetical surface reflectances. However, the counter is that the *layout* of surfaces *is* accessible (as in Figure 2.4). There must be information available to the perceiver that uniquely specifies that property.

In the acoustic realm, there are different ways to produce a given speech signal. Approximately the same formant pattern can be achieved either by rounding the lips, lowering the larynx, or doing a little of both. Quite different drum shapes—for example two different forms/shape/size of drums—can produce the same sound. — The "inverse problem," argues GLT—mapping speech signals onto vocal tract shapes that produced them—would have to be solved. However, a counter, as we have seen with Purves and Lotto (Chapter 3, [Purves2010]), is that the inverse problem is overrated, that there is sufficient information, and the problem is solved by pure statistical exposure.

BERGSON AND THE MOTOR THEORY OF COMPREHENSION

As noted, Bergson's analysis of the "storage of experience" began in the context of memory losses—aphasias—loss of understanding of language, words, specific words, classes of words. He had to set a basis in the very nature of language comprehension. He began his analysis as though he was sitting in a restaurant in France:

I listen to two people talking in a language that is unknown to me ... the vibrations that reach my ears are the same as those which strike theirs. Yet I perceive only a confused noise in which all sounds are alike … In this sonorous mass, however, the two interlocutors distinguish consonants, vowels and syllables which are not at all alike, in short, separate words. Between them and me where is the difference?" ([Bergson1896], p. 109)

We should step back here for a moment. This is a question on the origin and cause of a fundamental *phenomenal experience*. To be frank, it would seem that the charge toward and enthusiasm for the machine theory of mind, to include the ever-increasing perception of a succession of triumphs in the machine understanding of language, has caused us to have so "lost the plot" that this question has faded into non-existence. How does digitizing the acoustic stream (Bergson's "sonorous mass"), normalizing it, running it through a vast series of matrix transformations, and so on, account for the *experience* of this: *the clear perception of the ("separate") words of a spoken language?* Yes, one can take the standard AI stance that this "simulation" and understanding is not AI's problem, and that whatever method needed to achieve natural language processing is fine. But when this limited aim is lost, when claims are made that the brain is just doing a form of GPT3 processing but with less efficiency (for starters the poor deficient brain doesn't have the wherewithal to listen to 680,000 hours of speech while reading the written transcripts), this "losing of the plot" is on full display. So, this is the question:

How can the knowledge of a language, which is only memory, *modify the material content of a present perception*, and cause some listeners to hear what others in the same physical conditions do not hear? ([Bergson1896], p. 109, emphasis added)

Thus, shades here of what (re MT/DRT) we have just been reviewing:

...if auditory impressions organize nascent movements, capable of scanning the phrase which is heard and of emphasizing its *main articulations*. (p. 110, emphasis added)

Note the "main articulations." Bergson does not want or need (or think possible) absolute precision of consonant/vowel identification or segmentation. (We shall see why.) He does not need unique, precise specifications. He does not need to 100% solve the "inverse problem."

> These automatic movements of internal accompaniment, at first undecided or uncoordinated, might become more precise by repetition; they would end by sketching *a simplified figure* in which the listener would find, *in their main lines and principle directions,* the very movements [vocal "gestures"] of the speaker. (p. 111, emphasis added)

And:

> Thus would unfold itself in consciousness, under the form of nascent muscular sensations, the motor diagram as it were of the speech we hear. (p. 111)

So, to "adapt our hearing to a new language" would consist of:

> ...at the outset, neither in modifying the crude sound nor in supplementing the sounds with memories; it would be to coordinate the motor tendencies of the muscular apparatus of the voice to the impressions of the ear; it would be to perfect the motor accompaniment. (p. 111)

Now we shall again see something familiar: the *dynamic scheme* (Chapter 6, "Reaching for Cups—Voluntary Action"), that is, the principle of his 1902 *Revue Philosophique* article [Bergson1912] sketched out here in 1896 (Figure 7.4), but the implications for *acting* are not the more developed ones of 1902:

FIGURE 7.4. Learning the components of a tennis serve.

> In learning a physical exercise [a dance, a tennis serve], we begin by imitating the movement, as our eyes see from without. Our perception is confused; therefore, will be a movement whereby we try to repeat it. But whereas our perception was of a continuous whole, the movement whereby we endeavor to reconstruct the image is compound and made up of a multitude of muscular contractions and tensions ..." (p. 111)

Given the just preceding discussion re dynamic schemes (coordinating the components of a dance), pronouncing any *word*, given all its elements comprising the full action, would also involve a dynamic scheme. This will add complexity down the road. Continuing:

> The confused movement which copies the image is, then, already its virtual decomposition; it bears within itself so to speak, its own analysis. The progress which is brought about by repetition and practice consists merely in unfolding what was previously wrapped up, in bestowing [on] one of the elementary movements that autonomy which ensures precision, without, however, breaking up the solidarity with the others without which it is useless ... . The true effect of repetition is to decompose and recompose, and thus appeal to the intelligence of the body. (p. 111)

The diagram is not internal speech, and clinical facts show this: motor aphasics, who cannot pronounce words, can still understand.

> This is because the diagram, by means of which we divide up speech, indicates *only its salient outlines*. It is to speech itself what the rough sketch is to the finished picture ..." (p. 112, emphasis added)

In this, we need to distinguish between *production* and *comprehension*:

> For it is one thing to understand a difficult movement, another to be able to carry it out. To understand it, we need only to realize what is essential, just enough to distinguish it from all other possible movements. But to be able to carry it out, we must have brought our body to understand it. (p. 112)

Continuing:

> Now the logic of the body [like computer programming] allows of no tacit implications [no undeclared variables]. It demands that all the constituent parts of the movement be set forth one by one and put together again. Here a complete analysis is necessary ... the diagram is but a sketch. (p. 112)

So how can such an accompaniment be produced? To effectively pronounce a word, he notes, we know that:

- The tongue and lips must articulate.
- The larynx must be brought into play for phonation.
- The muscles of the chest must produce an expiration of air.

So, for every syllable uttered, Bergson notes, a number of mechanisms are already prepared in the cerebral and bulbar centers. These mechanisms join to the higher centers of the cortex via the pyramidal cells.

> Along this path the impulse of the will travels. So, when we desire to articulate this or that sound, *we transmit the order to act* to this or that group of motor mechanisms selected from them all. (p. 113, emphasis added)

The Auditory Aphasic Phenomena

Certain aphasic phenomena show these mechanisms (underlying the motor-diagram) are related to the perception of speech:

- A case of Lichtheim:
  - Due to a fall, the subject had lost the memory of how to articulate words—thus could not speak spontaneously. Yet ... he could repeat correctly what was said to him.
- Another Lichtheim case:
  - Where spontaneous speech is unaffected, but word deafness absolute, the patient not understanding what is said to him, he can still repeat what is said to him.
- Echolalia:
  - The subject repeats mechanically what he hears, as though auditory sensations convert automatically to articulations.

▪ Another case:

- The patient had retained both the auditive memory of words and the sense of hearing, but could understand nothing said to him.

Bergson notes:

> [All these] testify to a tendency of verbal auditory impressions to prolong themselves in movements of articulation; a tendency which assuredly does not escape, as a rule, the control of the will…and expresses itself…by an internal repetition of the striking features of the words that are heard. Now our motor diagram is nothing else. (p. 114)

We must remember, the crude auditory perception is really the continuity of sound, the "sonorous mass." The diagram—the sensorimotor connections established by habit—breaks up this continuity. A lesion to these (thus to the supporting apparatus for the diagram) hinders the decomposition and would check the upsurge of verbal memories "which alight on the perception."

Cases indicative of this motor diagram damage:

▪ Adler—in the word deafness—patients no longer react to even the loudest sounds, that is, the sound no longer has its motor echo.

▪ A patient of Charcot—with temporary word deafness—heard his clock strike but could not count the strokes. It is likely that he could not separate and distinguish them.

▪ Another patient—perceives the words of a conversation as a confused noise.

▪ Another patient—can't understand spoken words, but recovers if the word is repeated slowly, several times, with marked divisions, syllable by syllable.

Says Bergson: "All these facts combine to prove the existence of a motor tendency to separate sounds and establish their diagram" (p. 116).

Brief Stop at Vision

It is instructive to take the "motor-accompaniment" in the context of vision. We are all familiar with the experience of moving into a new city and experiencing the initial lack of a feeling of familiarity or recognition with our environment. However, after time, we become quite familiar—our body becomes automatically attuned to the environment—we and our body know exactly where we are, where to turn, where we are going, and so on. Just like walking down our driveway—there is a familiarity. This feeling arises from the development over time of an automatic *motor or action system accompaniment* to our perception. The street layouts, gradients, flow fields, building contours—all bring about an automatic organization of potential actions.

Say a certain lesion to the brain disrupts the mechanisms that create this schema of organized actions. We have the phenomenon of a standard "aphasia" disorder. We look at the street and no longer recognize it, nor do we have the slightest idea which way to turn. But has the memory of the streets, the city, the normal habitat been lost?

Bergson pointed out a case described by Wilbrand: The patient could describe with her eyes shut the town she lived in, and in imagination walk through its streets. Yet once in the street, she felt like a complete stranger; she recognized nothing and could not find her way. What is lost: a certain action accompaniment or organization built over time relative to the perceived spatial layout. What is *not* lost: the memory image, that is, the ability to redintegrate the experience (memory).

We are surrounded by objects of our everyday experience which are normally associated with certain sets of actions: spoons, forks, tables, chairs, bowls, hammers, bananas, and so on, all of which we have used repeatedly over time. Again, what occurs when a lesion to the brain disrupts the mechanisms which organize the normal motor responses to such objects? There occurs a standard "agnosia"—the failure to recognize objects. Thus, a victim who was once a writer may be shown a pencil and have not the slightest recognition of its function. Yet the patient can summon a mental picture of the object named and describe it very well.

These auto-accompaniments are now termed "action syntagms." The overt actions are eventually inhibited. But lesion the inhibitory networks, this results in another syndrome: we see a cup and cannot help ourselves from reaching for it to grasp it.

This "automatic motor accompaniment" has multiple components or processing phases, some of which are related to perceptual analysis, some to the grouping of relations, and some ultimately to the relation of perception to action. C.K. received severe trauma to the brain in an auto accident [Behrman1994]. He is very poor at recognizing visually presented objects yet can name and recognize the same object presented to the touch. His visual recognition failures appear to stem from an inability to segment and group elements in an array. He cannot match or trace items that are overlapping and cannot segment a figure from the ground as visual noise is increased. He can pick up local features of an object but not organize them as a whole: A dart is called a feather duster. A tennis racket is called a fencer's mask.

C.K. copies geometric forms in a slavish, piecemeal fashion without appreciating the identity of the object being copied. Yet despite this recognition deficit, his mental imagery, and thus his memory images of objects, is intact. He can draw wonderfully in detail from memory the very objects he cannot recognize, and he can describe in detail the size, color and shape of letters and objects he fails to identify visually.

Piaget (*The Child's Conception of Space*, [Piaget1949]) described the development of the child's ability to reproduce a geometric form coherently, in terms of the development of mathematical group relations underlying the actions, that is, this is a complex ability that a lesion is affecting. Cassirer too – recall the aphasic who could not draw a room unless an "X" is drawn of the paper, showing his position (Chapter 4).

This was part of Bergson's argument regarding the storage of experience: the loss of certain memories, like what a pencil does, does not mean that experience is stored in the brain.

Learning the Motor Diagram

So, there is an automatic, a motor basis, for both visual recognition and for words; destruction of it does not mean memories, experiences are lost. Here is one more quick example of the "not lost." Bergson noted Ribot's observation (Ribot's law): In some amnesias, word loss gradually proceeds in this order: from proper nouns to nouns to verbs. Recovery is in reverse order: verbs to nouns to proper nouns. With this, firstly, there is the fact of yet another phenomenon where there is not actual memory loss at all. Second, why this order? Verbs imply simple actions, for example, stirring. The body can most easily be moved into a "motor attitude" to allow us to recover the word, that is, to redintegrate it, to retrieve experiences (like coffee stirring) in which words ("stir") are an intrinsic part.[1] Nouns imply objects that are a nexus of sets of actions—a piece of cheese—slice it, eat it, grate it, wrap it—thus more complex to redintegrate. Pronouns imply a person. This implies an even more complex set of actions + context/ environments (work vs. home vs. coffee shop).

So, again, there is an automatic, a motor basis, for visual recognition and for words (the auto-motor accompaniment). Underlying this, then, there are two forms of recognition:

- automatic: that issuing from the motor accompaniment (from the object)

- attentive: that reliant on the *complete circuit* formed with an *active memory* "projecting" memories (from the subject)

Implicit here (well, explicit in Bergson) is a model of *conscious* attention. Note the relation to explicit memory and the problem of the symbolic: We see the wind chimes on the porch ⇒ the past memory ⇒ a present for wife. The object is attentively recognized, permeated with memory, and symbolic of a past event.

This elementary distinction—automatic recognition versus attentive recognition (with "familiarity" a function of an automatic accompaniment to the environment) —to our knowledge is totally absent in cognitive science. Why? The automatic accompaniment here is a bodily *feeling*, an *intrinsic intentionality* as Crockett put it [Crockett1994]. There is no room for this in the pure symbolic architecture of cognitive science or AI.

---

[1] See Asher's studies on second language learning, where commands ("Stir," "Lift the spoon"), similar to the SPTs, are coordinated with performing physical actions, and equally a model for the command-permeated first language learning environment of the child ("Johnny, wipe your nose!"), that is, words (here, commands) are intrinsically part of events [Asher1965], [Asher1966], [Asher1967], [Asher1972].

All this is a precursor to the second part of the language comprehension story:

> These inner movements ... are like a prelude to voluntary attention. They mark the limit between the voluntary and the automatic. By them ... the characteristic phenomena of intellectual recognition are first prepared ... But what is this *complete and fully conscious recognition*? (p. 116, emphasis added)

In other words, we are about to bring in knowledge, in fact, the *semantics* involved in one of these initially, (somewhat) parsed by the motor diagram, acoustic streams, that is, sentences.

MEANING AND GENERATIVE AI

As we are in a compare-and-contrast mode, we want to look here first at what "meaning" is in the language processing model of current, generative AI and its transformer/attention architecture. The insight into the creation of this architecture is credited to Vaswani et al. and their "Attention Is All You Need" article of 2017 [Vaswani2017]. There are innumerable videos explaining this architecture (e.g., [Karpathy2023], [Manning2021a], [Manning2021b], [Ark2020], [Stat2023] [Svensson2021]) and a detailed tutorial is not intended here, but, especially for anyone uninitiated in this subject, there are a few foundational features to describe.

Words as Vectors

The critical, starting feature of the model, already well established before 2017, is that words are represented as vectors of numbers. The number-elements of a vector for a given word are probability values expressing the likelihood of that word being related to every other word in the "corpus" of text(s) available to (or trained on by) the system. Figure 7.5 gives a beginning idea of the concept of various mini-texts implicitly containing a word's probability of being related to other words. In this case, we are focusing on the word "banks" and words tending to "surround" or provide "context" for banks.

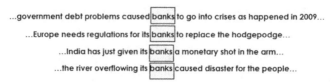

FIGURE 7.5. The word "banks" and sentences with "surrounding" context words.

If the system's total vocabulary is 500 words, there will be 500 word-vectors, each of length 500, where these vectors relate to the word's role as a *focus* word or *center word* (a "center" being the "banks" example). In the vector representing the first word (Word 1) in the vocabulary corpus (as a center word), the first element expresses the probability of the word being found with itself; this is always a 1. The second element is the probability of being found with Word 2, the third element is the probability of being found with Word 3, and so on, until the 500th element which is Word 1's probability of being near Word 500. Each of the 500 words has such a vector. There will be a second set of vectors, again one for each word, relating to its role as a *context word*. Now bear in mind that even with GPT-2 we already were dealing with a vocabulary of 50,000 words, while GPT-3 is on the order of 15 million words (though this includes multiple languages, not just English).

The treatment of words as vectors leverages on the concept of *vector spaces* and the associated concept of a dot-product of two vectors. In the simple 2-dimensional case of Figure 7.6 (not the 500-dimensional space that would be required for the 500-word vocabulary), the vector for "bank" is very close to the vector for "money" and farther away from the vector for "water." Note that, appropriately, the dot product of the bank vector with the money vector is larger than the dot product of the bank vector with the water vector. In other words, the larger the vector dot product, the closer the vectors are in the vector space, and the more closely related. One can intuitively see that if we want "water" closer to "bank," which it must be for comprehending the meaning of that last sentence of Figure 7.5, where the river overflows its banks, then, for that sentence, we will need to adjust values somehow to move those two vectors (water, bank) closer together in our vocabulary vector space. This will be the essential task of the "attention" algorithms.

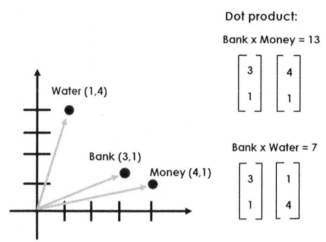

FIGURE 7.6. Word vectors in a (2D) vector space.

The algorithmic method to create these word-probability vectors is "Word2vec," described by Mikolov et al., in 2013 [Mikolov2013]. Looking at the essence of it adds more insight re the nature of this approach:

- There is a large corpus of text.
- Every word in a fixed vocabulary (e.g., 20,000 words of vocabulary) is represented as a vector.
- Go through each position, t, in the text which has a *center word*, c, and *outside* (or surrounding) *words*, o.
- A "window" is specified for the extent of the words surrounding the focus or center word, c, to examine, for example, five words on each side of c.
- Use the similarity of the word vectors for c and o (the dot product is a measure of this similarity) to calculate the probability of each o given c (or vice versa).
- Keep adjusting the word vectors to maximize this probability.

A schema of this process is shown in Figure 7.7. Here a small window size of 2 is being used around the center word, "banks," and the process is computing the probability, P, of the outside word occurring (say, at position -2, left of the center word, or w_{t-2}), given the center word at position t (w_t). Again, this is done across the entire text.

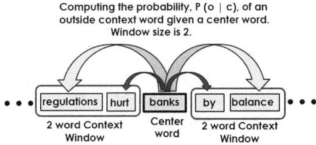

FIGURE 7.7. The sliding window process across a text for computing the probabilities of words occurring with some focus word (also known as "skip-gram"). (Adapted from [Manning2021].)

When all is said and done, we have a huge, multidimensional vector space, where the vectors express how the words cluster together (or, conversely, are far away). Projected on a two-dimensional surface it might look something like Figure 7.8 where we see clusters of related words. This is the *knowledge base* of the Generative Pretrained Transformer (the GPT).

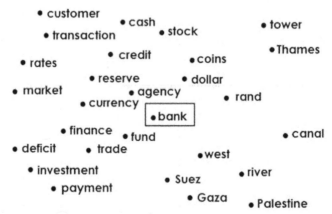

FIGURE 7.8. A 2-D projection of the vector space: clusters of probability related words.

The Attention Mechanism

It was a bit premature to say, "This is the knowledge base," as previously stated. There is an important process that modulates this knowledge, namely, "attention." In Figure 7.9, the algorithm for attention is pretty well summed up. We are considering an input sentence and, in the example, this is, "He swam the river to reach the other bank." If we keep in mind the clusters of Figure 7.8, "bank" is far more related to financial things than it is to water/river things. This is what the vector space represents, and therefore what vector 1 thru vector 9, representing each word in the sentence, will reflect. The vectors, in other words, are pointing in the direction of finance. This state-of-affairs is not good for comprehending what "bank" means in this "swam the river" sentence.

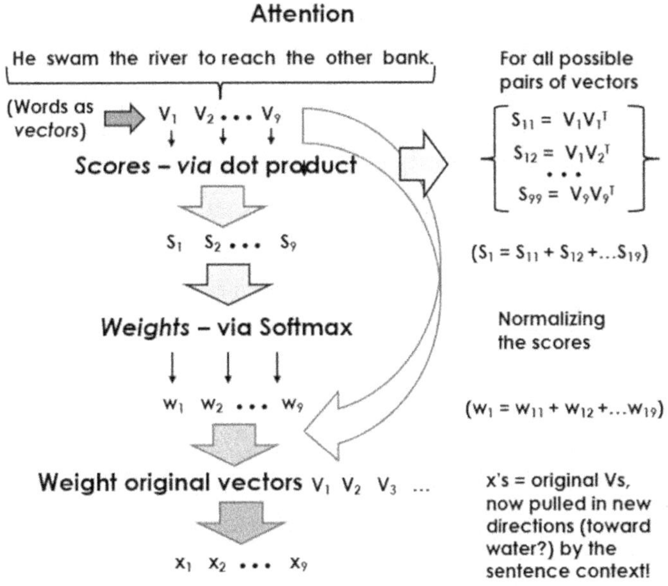

Attention

FIGURE 7.9. The "attention" algorithm.

The attention algorithm flow (Figure 7.9) first takes the dot product of the word vectors (nine of them) in the nine-word sentence, taking all possible pairs of vectors. These become "scores" which, in the next step, are turned into *weights* by being run through the SoftMax function, normalizing the scores (setting the values between 0 and 1). Finally, last step, the original vectors for each sentence word are adjusted by these weights (becoming the "*x*" vectors, i.e., the now weighted original V vectors).

In effect, what we have just done is adjust the original vectors based on the context of the sentence, where the sentence has words around water—river, swam. The original vectors are now, hopefully, pulled and pointing more to this other area of the vector space, this, shall we say, being now an important piece of knowledge for interpreting that sentence. If ChatGPT is shown the "swam the river" sentence, then asked, "Was the man swimming to get his money out the bank?" it will answer, "No, the word 'bank' refers to the bank of a river, not a financial institution."

LINGUISTIC UNDERSTANDING AND THE GPT

This is the foundational basis of generative AI. It is a basis that has turned out to be extremely powerful.[2] It has allowed AI to handle classes of sentences that involve the word-reference problems in speech that have been heavily resistant to solution, generally deemed to require a true knowledge of the physical world for understanding:

- The chicken crossed the road because it was hungry. [Who was hungry?]

- Winograd schemas, for example, Sam tried to paint a picture of shepherds with sheep, but they ended up looking more like golfers. [Who looked like golfers? The shepherds? The sheep?]

Certainly, there is far more capability now than just this disambiguation ability, arising as it does from the *generative* side of these systems which relies on the same knowledge base to generate sequences, from writing computer program code to generating art to instructions to create a mousetrap. The difficulty here is that this is raw, 18th-century *associationism* run wild. Due to the availability of massive amounts of data, which is to say, an enormous treasure of human knowledge already encoded in linguistic text, it has already gone farther than any early associationist or critic thereof quite dreamed.

The essence of associationism is to view the ecological world as consisting of *separate elements* that must be "associated" or become associated over experience. The connectionist neural networks already simply instantiated this. Take a tiny hypothetical

---

[2] We are neglecting, among *many* other things, details like positional information, multiheaded attention, the important role played by the multiple indexes established into these vectors, and as well, the indexing to entire text documents.

section of the neural network described by Rogers and McClelland [Rogers2004], where we presume it has been trained to associate functions/actions with "spoon" via the attribute of what the spoon "can do" (Figure 7.10), or similarly, to associate aspects of "coffee stirring" via the attribute "has."

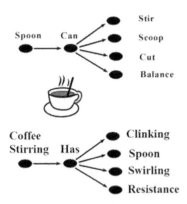

FIGURE 7.10. A schematic connectionist net associating the elements of coffee stirring.

What sense does this make? What sense does it make to laboriously train a network over many trials, to back propagate via derivatives to adjust those weights, just to get the net to respond, "Coffee stirring *has*: [swirls, cream, clinking, spoons...]"? These "elements" of coffee stirring are already intrinsic, integral "parts" of the event, *avoidably so by the very physics of the event* [Robbins2008]. An instrument (like a spoon) is *necessary* to move the liquid, a *force* is needed to move the spoon. A swirling, radial flow field is naturally produced by the spoon's force and motion. A "clinking" is intrinsic as the spoon glances against the cup's side. A constraining container (like a cup) for the liquid is intrinsic, necessary, or the liquid flows all over the tabletop. A resistance to the force/motion inherent in the liquid is there. All of this—or "coffee stirring" just does not happen. *None of these "elements" need to be "associated."* They do have to be *differentiated.* From out of this whole, continuous, intrinsically self-related event, they must be identified, isolated, but not associated. And these are just the elements we have bothered to explicitly identify. There is the kinesthetic flow field as we stir the spoon with its adiabatic invariance defined over the spoon's periodic oscillation, that is, the various feelings of stirring; there is the swirl of the liquid with no particular name, the peculiar mixing of cream into the brown coffee, the feel of forces involved in that inertial tensor which also carries (specifies) the constancy of the spoon's length and size as we move the spoon,

None of these *associated,* static "elements" (again, with no *principled* derivation) have anything to do with the *invariance structure* of the coffee stirring event; they do not express it; they do not come close to capturing it. Associated elements are *not* our experience; they do not capture the invariance across the 4-D memory (or "stack") of these dynamic experiences which is our concept of stirring coffee. Yet it is obvious that the association structure which comprises the base "knowledge" of generative AI is only another variant of this connectionist associationism, a conception which at its core makes no sense.

GENERATIVE AI AND THE DESTRUCTION OF CHOMSKY

Simply because we have displayed a syntactic rule structure a la Chomsky (in the context of Lashley, Chapter 6), it is hopefully not presumed that all aspects of Chomsky's approach to language are being endorsed here. The fact that we have put the atemporal (or contemporal) idea at the apex of this rule structure—an idea necessarily involving the invariance structure of the event about to be expressed via elements of language—is denoting that far more is going on when expressing an idea according to the "rules of linguistic syntax" than Chomsky acknowledged or envisioned, in fact, as we follow the unfolding of Bergson's view, something very, very different is the case. It is interesting therefore to see generative AI being credited now with the destruction of Chomsky and generative linguistics, a claim whose basis we think can only be very, very partially correct.

Piantadosi [Piantadosi2023], in his very positive assessment of the generative pretrained models, has made this argument, namely, that these models have undermined Chomsky, in fact, more critically, that these models have replaced Chomsky and are now, by default, *the new basis of a theory of human language understanding*. He notes two main features of the models as,

(1) an *attention mechanism* (allowing the next word in sequence to be predicted from the previous far in the past), and (2) the *integration of semantics and syntax*. In addition, there is the prime architectural feature where internal representations of words in these models are stored in a vector space, to include properties that determine how words can occur in sequence (e.g., syntax).

Some of the previous discussion (attention, words as vectors, vector spaces) refers to what we have already described with respect to the generative AI model. We have a fairly good idea of what these two features consist of, that is, we know what kind of architecture upon which Piantadosi is basing his argument. The essence of the argument that these models replace any mechanisms Chomsky envisioned for the acquisition of syntax (but remain unimplemented, unproven by Chomsky and generative linguistics) is that these GPT models have demonstrated such a mastery of syntax that they now must be seen as the new basis for human language acquisition, and thus, by implication, what the brain does. To include, one can ask, brain-storage of words as vectors of probabilities?

Whether these statistical models account for the acquisition of syntax is not going to be the question here. Misgivings on this might be noted: Arguments that the models account for 70% (or more) of correct performance on problematic structures (like the Winograd schemas) have a very arguable value as actual proof that this is what the brain is doing. They unfortunately sound like Newell and Simon's argument that their GPS accounted for a good percentage of the verbal traces of humans doing problem solving (the human "protocols" recorded by Newell and Simon) on the same tasks (like cryptarithmetic, theorem proving). The question is whether humans were doing anything like the GPS algorithm (or not) despite the percentage of match being claimed to the GPS steps. But it is the claim that the models are capturing semantics that is the worse problem. Thus, he notes that the language model vectors appear to encode at least some aspects of semantics, with these "… building on earlier models that encoded semantics in neural networks [Rogers2004] [Elman2004]" ([Piantadosi2023], p. 12).

And further on, an even more positive assessment, for though current versions of the models achieve their performance via very large network sizes and large amounts of data, he feels that it has been shown (by [Zhang2020]) that actually most of this learning is not about syntax, but rather, that models trained on the order of 10–100 million words, "… reliably encode most syntactic and semantic features of language and the remainder seems to target other skills (like knowledge of the world)" ([Piantadosi2023], p. 14).

But from the perspective of this work, there is absolutely zero semantics in these models. There exist in these models nothing like the invariance defined across the concrete events of our 4-D experience—the coffee stirrings, hammer and nail poundings, napkin foldings—and the redintegrative wave passing through this experience that defines these concepts. The associationist network of Rogers and McClelland referenced by Piantadosi as "encoding semantics" [Rogers2004], does no such thing (something we have previously noted regarding this model as basically non-sensical, along with problems also discussed in Chapter 4). Words as probability vectors do no such thing.

One's intuition is that these invariance structures defined over concrete events are so rich, so full of details never spoken of in language, so vast, that these generative AI models simply must ultimately be tripped up, failing on something obvious to a human. The problem is twofold: (a) the language corpus derived from human experience that these models access is now so vast, it is very difficult to come up with these cases, in purely the linguistic realm, that will catch a GPT, and (b) even if one does come up with cases, the AI theorist will simply say, "Oh, that's just part of the 2% (or whatever percentage claimed) of concrete world cases we can't *yet* handle— *yet*." In other words, the exercise is almost to no avail.[3]

So, we settle on one question here for AI: *What if there were no language?* Again, throw us back to the Stone Age: Are we saying we could not fashion a stone knife from flint and antler bone? Or that lacking a spoon for my stone age coffee cup made of stone, we could not fashion or find a substitute, like a handy stick (to maintain the invariance structure required for and defining coffee stirring)? That we could not track a deer? Make a bow and arrow? In other

---

[3.] See Yudkowsky admitting the same problem regarding the vastness of available text, here in the context of linguistic probes of GPT-4 as to whether it has awareness ([Yud2023], at 3.00).

words, that we could not exhibit, quite perfectly, ecological general intelligence? Are we really going to hold this? But given the absolute dependence on word vectors, this is exactly, inescapably, what is being implied.

This is what is implied when Chalmers [Chalmers2022] extols the SayCan robot project, aimed as it is at having a GPT drive the robot action, (e.g., "Insert the razorblade into the pencil") giving the appearance of what will then be termed "ecological" action. As we mentioned already in Chapter 1 and hit also in the Roli et al. "the robot and the engine block" discussion of Chapter 5, place the GPT-driven robot too in the Stone Age, the internet and its word-vector database removed; now ask it to construct a stone age knife. These models of intelligence start with words. Words are their entire basis. Nothing else. One is tempted say, "Live by the word, die by the word." This absolute dependence on pre-distilled human thought in the form of language simply cannot be the case for the human mind.

COMPREHENSION AND THE WHOLE OF MIND

A linguistic string, "The man stirred the coffee with the spoon," might be better seen as a mediating device to modulate the redintegration of a specified event with all the invariance laws implied and contained in our 4-D experience. Equally so for, "The man swam the river to get to the other bank." DiLeep George, who, it is safe to say, is dissatisfied with the current Generative AI approach to language [George2023], approaches this concept, noting that, "Language is for creating simulations in peoples' minds." The knowledge for the simulations, he argues, resides in the "perceptual-motor" system. If we take the sentence, "John pounded the nail on the wall," all kinds of questions of this sort can be asked:

- Was the nail vertical or horizontal?
- What was the sound like?
- What did the nail feel like?
- Can you smash your thumb?

To answer these questions, he notes, we mentally simulate the pounding of a nail, and from this we pull the answer to the questions.

But this is in effect an appeal to (a) the invariance structure of events (like pounding nails) existent across our 4-D experience, and (b) an implicit appeal, if still in the standard cognitive science/AI framework, to what is and will be an eternally elusive theory (for AI) of the origin of mental imagery.

FIGURE 7.11. A biting transformation: a series of images as "states."

In regard to (b) and imagery, George appeals to Barsalou's theory, focused as it is on the importance of the image (the image as "perceptual symbol system") [Barsalou1993], [Barsalou1999]. Barsalou himself failed to note the Gibsonian significance of the dynamics of these events. When considering how he would store an event like "biting a carrot," he proposed storing three *states* as images: a mouth closed, a mouth opening, a mouth closed around the carrot (Figure 7.11). The difficulty here is that "biting" is a facial flow field, just as dynamic a transformation over an indivisible flow as the swirling and forces of coffee stirring or the rotating cube. (Cf. Google images, "hamster biting a carrot" gif.) It cannot be stored as a series of discrete states, and certainly not as "images" in the brain. Dileep George, we would propose, were he to make deeper considerations, could find himself working within the Bergson-Gibson framework.

In this comprehension process, the whole mind is present, is brought to bear, the entirety of experience. The 4-D extent of our experience is seamless, without breaks, without discrete elements. But it is focused, dynamically lensed, by that modulating linguistic string to support a specific meaning. An interlocutor speaks to an experienced canoeist, someone who has experienced many canoe trips—the lakes, the waves, catching lake trout, walleyes, frying fish for breakfast, setting up tents, campfires, wood chopping, axes, storms, portages, swamps, bears, and so on. With the spoken sentence, "The man paddled the canoe," the whole of the hearer's being is focused—all those experiences, all his experience. This underlies the meaning of that sentence. A hint of this is given in the LSD experience where the biochemistry underlying the normal, "tight" organization of various centers of the brain and the close tie of perception and action is being loosened, and past experience is flooding into awareness through Bergson's metaphorical brain-as-a-valve, a valve normally set to only let very relevant-to-the-moment experience be accessed (redintegrated). Such a simple sentence now seems to "reverberate" with unending meanings, precisely because, we would hypothesize, the whole of the mind and all experience is in reality behind the understanding of the sentence.

In fact, the whole of the mind is even behind the parsing of that continuous acoustic stream, in which the linguistic string is "embedded," into the separate words which we perceive to be there—man, paddled, canoe. This brings us back to the perception of that speech stream, now in the context of our comprehension of that sentence.

AN *ACTIVE* MEMORY

So far, we have been portraying the 4-D extent of being as, in essence, static, as just a trailing extension (of our being) in time (as in Figure 5.9). But this "extent" is dynamic, active. One can ask, why should it not be (active)? It is all "me." A small extent in time comprised our entire active being during an event, for example, stirring breakfast coffee, but we can ask a form of Gibson's question, "Where is the dividing line?" Where is the point at which that extent—our being—becomes "static?" There is no such point.

So, Bergson introduces what we'll term "Part 2" on linguistic comprehension as such:

> We have said attentive recognition is a kind of circuit in which the external object yields to us deeper and deeper parts of itself, as our memory adopts a correspondingly higher degree of tension in order to project recollections towards it. ([Bergson1896], p. 116)

So, note the concept of "tension" and active "projection." We shall have to explore this in a bit. He continues:

> In the particular case we are now considering, the object is an interlocutor whose ideas develop in his consciousness into auditory representations which are then materialized and uttered as words. (p. 116)

And:

> So, if we are right, the hearer places himself at once in the midst of the corresponding ideas, and then develops them into acoustic memories which go out to overlie the crude sounds perceived, while fitting themselves into the motor diagram. (p. 116)

Remember: we have had no theory of "S" —Lashley's contemporal idea—from which everything—the syntactic structure of the sentence—devolves. So, noting that to understand another's words is *to reconstruct, starting from the ideas, the continuity of sound perceived*:

> And more generally, to attend, to recognize intellectually … may be summed up in a single operation whereby the mind, having chosen its level, having selected within itself, with reference to the crude perceptions, the point that is exactly symmetrical with their more or less immediate cause, allows to flow towards them the memories which overlie them." (p. 117)

"Filling In"

There are simple examples of this memory "projection." We hear a new song on the radio, but the words are very hard to understand, to catch. We cheat—someone tells us the words, or we find the text, or even just receive an explanation of the context. Now the words are heard distinctly, clear as a bell. Or another: We are walking in the woods, and suddenly we see a "bear" along the trail. Creeping a bit closer—it's just a tree stump. The past memory was "projected" within the perception. In this vein, Bergson noted studies on reading. These showed that very few letters are actually seen. In his words, reading is, "a real work of divination" ([Bergson1896], p. 126).

In recent years, this pretty much went under the notion of "filling in" (Cf. [Dennett1992]). For example, the brain, rather than "seeing" each individual tile on a wall, is "filling in" the tile pattern. A missing tile may not even be noticed. This is construed, usually, as though the whole image—the entire wall—is simply *constructed* by the brain. This goes well with the thesis that "the world is an illusion" and of course, indirect realism, where the brain simply, but mysteriously, constructs the image of the coffee stirring, that is, phenomenal experience. Unfortunately, what is needed is a model of the origin of the image of the external world in the first place, before we can even have a nice set of images stored in the brain (somewhere, somehow) available to use as "fill in," or, more usually, that these images are simply "generated" (with no theory, save "emergence") as fill-ins along with rest of the generated image of the external world. The entire "filling in" discussion, to this point, has been in this sense a bit nonsensical. It has presumed that a model of the origin of the image would not change the entire discussion radically.

So, Bergson notes:

> We *start from the idea,* and we develop it into auditory memory-images capable of inserting themselves in the motor diagram, so as to overlie the sounds we hear. ([Bergson1896], p.122, emphasis added)

But cognitive science is unwilling to go this way:

> .... because of the invincible tendency which impels us to think on all occasions of *things* rather than of movements. (p. 121, emphasis added)

In other words, the classic metaphysic of space and time holds sway in all theory. But here is yet another, "Where is the dividing line?" question that this metaphysic forces:

> We have a continuous movement ... . At no moment is it possible to say with precision that the idea or memory-image or the sensation begins. And in fact, *where is the dividing line between the confusion of sounds perceived in the lump and the clearness which the remembered auditory images add to them* ...? (p. 121, emphasis added)

Continuing:

> ... or between the discontinuity of these remembered images [the separate words] and the continuity of the original idea [e.g., stirring the coffee] which they *dissociate and refract into distinct words*? (p.122, emphasis added)

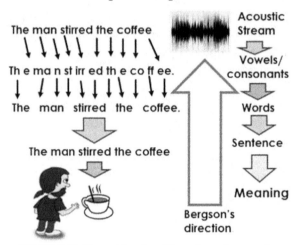

FIGURE 7.12. The unidirectional flow of speech processing (versus Bergson's direction: from the idea toward the motor-diagram).

Rather, we get the picture of Figure 7.12, namely, a series of discrete stages, where we start with discretizing the acoustic stream into separate elements. From this stage we move on to the next, where we assign meaning to the elements. In other words, we have a unidirectional flow from the acoustic stream, to elements, to meaning.

> [Cognitive theory] erects the crude sounds heard into separate and complete words, then the remembered auditory images into entities independent of the idea they develop: these three terms, crude perception, auditory image, and idea, are thus made into distinct wholes of which each is supposed to be self-sufficing. (p.122)

And he notes: "We [cognitive theory] see no harm in reversing the real order of the processes, and in asserting that we go from the perception to the memories, and from the memories to the idea" (p.122).

Yes, we have painted an over-stark, over unidirectional picture of the AI process here, for a certain amount of "semantic" processing, some context, is used to help disambiguate the words in the acoustic speech stream when we talk, say, to SIRI. But this is a mere inconvenience, a bothersome complexity in what otherwise would gladly be taken as a pure, unidirectional process from speech stream, to words, to meaning.

The Storage of the Elements (Words)

All the components of these stages are considered stored in various areas of the brain, from which areas, lines of communication, connections, run everywhere:

> Nevertheless, we cannot help feeling that we must bring back again, under one form or another, the continuity we have broken ... we go on materializing this [continuous] development itself into lines of communication, contacts, and impulsions. (pp. 122–123)

Each new fact, each new lesion, each new memory malady will force us to complicate the diagram of centers and connections, to add new stations, centers, and so on, "Yet all these stations, laid out side by side will never be able to reconstitute the movement itself" (p.123).

Bergson noted the diagrams, with centers of "visual" memories, "auditory" memories, "tactile" memories. The centers multiplied. But these maps were all eventually abandoned. He noted, it's as though theorists *never considered the structure of a sentence*, that is, everything required to help express a dynamic event:

> They argue as if a sentence were composed of nouns which call up images of things. What becomes of parts of speech, of which the precise function is to establish, between images, relations and shades of meaning of every kind. (p.124)

Thus, any concept that the brain has some area for the storage of words collapses under the weight of phenomena truly being considered:

> Now consider for a moment the amazing consequences of a hypothesis of this kind. The auditory image of a word is not an object with well-defined outlines, for the same word

pronounced by different voices, or by the same voice on different notes, gives a different sound. So … you must assume that there are as many auditory images of the same word as there are pitches of sound and qualities of voice. Do you mean all these images are treasured up in the brain? (p. 118)

He goes on here, enumerating a very long list of logical difficulties, and this is where the clinical facts (already noted, "auditory aphasic phenomena") also take issue with any storage of words hypothesis. Certainly, the AI endeavor hit the essence of this problem long ago; this is why the sounds of the acoustic stream are normalized, but the drive is to turn this into those tokens—man, swam, river, bank—thus the vectors, then operate with this against the 50,000 word-vector vocabulary. This is just word-storage in a form Bergson could not have visualized (and to which Lashley would happily have taken his surgical knife).

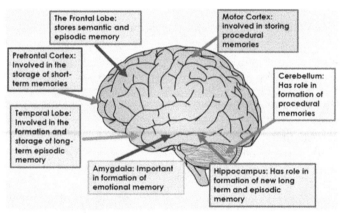

FIGURE 7.13. A current view of memory storage.

Poeppel and Idsardi, in an article entitled, "We Don't Know How the Brain Stores Anything, Let Alone Words" [Poeppel2022], number among the few calling attention to the problem. It is quite curious today: Scan through, "memory storage areas in the brain" on Google images. One will not find anymore (Figure 7.13) actual labeled storage areas, like "visual memories," "auditory memories," "words," "experiences." Rather, one finds, "involved in the storage of," "important for the transfer into long-term memory (LTM), or purely "functions" (processing). And LTMs are stored in "multiple areas." Fuster [Fuster1994] had indicated, as earlier noted (Chapter 4), the cause behind this, namely that "all 'storage' areas were [discovered to be] simultaneously *processing* areas," which is to say, there may well be no storage areas. Again, this is simply the question of Chapter 2 when looking at the dynamics of events like "coffee stirring" or the strobed, rotating cube, and the brain resonating to the ongoing external event-transformation with all the feedback and reentrancy involved, namely, at what point(s) in time would the brain be able to stop and "store" anything in this process. Or just when would the hippocampus, clearly involved in the processing of the events being perceived, begin, or start *driving* (as it is held to do) its "consolidation" process, acting as the "index" for storing the event in those multiple, various cortical spots as the event is ongoing? In truth, Bergson quite accurately described this state-of-affairs as near a fait accompli in 1896 and forecasted the state we see today. That is to say, this is another elephant in the room: where, precisely, are experiences stored, or words stored? There is silence (unless we believe they are probability vectors). So, he stated: "We are bound to consider what becomes of the known facts when we cease to regard the brain as a storehouse of memories" (p. 128).

Some Form of Internal Keyboard

Bergson notes that every perception gives rise to a number of sensations, all coexisting and arranged in a certain order. In the case of an external object, like the coffee cup, this order comes from an "organ of sense." The initial visual area, V1, had not been described at the time, but this would be that "organ of sense" for a visual object.

> It is like an immense keyboard, on which the external object executes at once its harmony of a thousand notes, thus calling forth a definite order, at a single moment, a great multitude of elementary sensations corresponding to all points of the sensory centers that are concerned. (p.128)

It is from this physiological area, in the Gibson framework, that the resonance process begins, specifying the external object. The keyboard area is dynamically changing, down to the most minute interval of time. Gibson, would simply disregard (take very lightly) Bergson's use of the term "sensations" here, considering this merely a useful turn of phrase to express the reality happening—the *information* arriving—at the "keyboard." Bergson notes:

> Suppress or remove the keyboard—the strings are still there—but there is nothing to activate them, to let a thousand notes be struck at once. (p. 128)

Now here is where Bergson's model introduces a difficult question:

> In our opinion, the "region of images," if it exists, can only be a keyboard of this nature. Certainly, it is in no way inconceivable that a purely psychical cause should directly put into action all the strings concerned. ([Bergson1896], p. 129)

The psychical cause acting on the physical keyboard is "causally" problematic. We must look further at the implications of the temporal metaphysic to make sense of this. For now, we continue down the intuitional path Bergson is being led by looking at the clinical facts— facts which simply cannot be treated as non-existent by AI theorists purporting that such systems are explaining the brain. Thus, for hearing, he notes that the localization of this function appears certain: a certain definite injury of the temporal lobe destroys it. When factoring in the many reasons for the impossibility of storage of images:

> Hence only one plausible hypothesis remains, namely that this region [an area of the temporal lobe] occupies regarding the center of hearing itself that place that is exactly symmetrical with the organ of sense. It is, in this case, a mental ear. (p. 129)

In essence, these initial keyboards may be activated from two directions: from without, via signals from the sense organs, that is, by a real object, and from within, through successive intermediaries, by a virtual object. So, the auditory image, called back by memory sets in motion the same neural elements as the original perception (that is, the elemental articulations comprising the motor diagram). And the recollection must change gradually into perception. Thus, recalling to memory complex sounds (words) can concern different aspects of the neural system as opposed to the mechanisms involved in the initial perceiving.

So, in *psychic deafness*, one can still hear physical sounds—the acoustic stream—but not hear the separate "words" (as the memory images cannot actualize). "The strings are still there, and to the influence of external sounds they vibrate still. It is the internal keyboard that is lacking" (p.129). This, he notes, is a simplified picture: The various aphasic syndromes show that calling up an auditory image is not a single act. Between the *intention* (the pure memory) and the auditory memory image there can be other memories realized as memory images in more or less distant centers. It is by successive degrees that the idea comes to embody itself in the verbal image(s) per se. So *mental hearing* depends on the integrity of various centers and paths.

> Whatever the number of intervening processes, we do not go from the perception to the idea, *but from the idea to the perception*; the essential process of recognition is not centripetal, but centrifugal. (p.130, emphasis added)

Given understanding even the simple, "The man stirred the coffee with the spoon," the sources of complexity here are multiple:

- the multimodal stirring event (idea) itself with its invariance structure
- the multiple individual word images: the, man, stirred, coffee, with, spoon
- the relations established, for example, a man doing the stirring, using a spoon
- the linguistic syntax that must be respected

But this is the problem already mentioned, which we will visit later:

> Here, indeed, the question arises how stimulation from within can give birth to sensations, either by its action on the cerebral cortex or other centers. But it is clear enough that *we have here only a convenient way of expressing ourselves*. (p. 130, emphasis added)

However, subsequently, Bergson does more of a recap, defining the nature of voluntary action (of which speech is also a case).

> Pure memories, as they become actual, tend to bring about, within the body, all the corresponding sensations.

But these virtual sensations, in order to become real, must tend to urge the body to action … The modifications in the centers called sensory, modifications which usually precede movements accomplished or sketched out by the body [a reference here to perception as virtual action] … are less the real cause of the sensation than the mark of its power and the consolidation of its efficacy. (p. 130)

And:

The progress by which the virtual image realizes itself is nothing less than the series of stages by which this image gradually obtains from the body useful actions or useful attitudes. (p.130)

MT/DRT AND BERGSON

At this point, what have we seen? This model is far from nonproblematic, far from simple, yet keep in mind a couple of things: (a) AI models are making no attempt to replicate human language perception, and (b) the current human models are (at least pretty much) very stalled (though Turvey [Turvey2023] (p. 190) feels the phonetic perception problem can be solved). Once we have an actual model of perception—the origin of the image of the external world—we have to follow its consequences.

Bergson's "motor diagram" looks to side with MT. We cannot construct such a diagram, even if only a sketch of vocal movements, without some form of invoking the neuromotor commands, else there is no "sketch." But he would equally agree with DRT: There is acoustic information specifying the source of the sounds, and we've seen that he would totally agree in the case of vision. But this sound information is likely insufficient for *unique* specification.

For Bergson, information is so only because related to the body's possible action, that is, perception is virtual action, and for the speech stream, this is the motor diagram. Gibsonians, despite "affordances," have never quite assimilated this. Both MT and DRT (and GLT) are industriously trying to show how things parsed or identified in the stream are *unique*, discrete elements (vowels, consonants), that is, are unambiguously specific. From these we have, again, a bottom-up build: syllables ⇒ words ⇒ sentences ⇒ meaning. For Bergson, this discrete starting point (vowels/consonants) is a lost cause. We have only a *sketch* of sound production. We need memory!

THE VIRTUAL

We have been showing images of events (like coffee stirrings) in 4-D memory (Figure 5.9). But, per Bergson, this virtual extension of being—"pure memory" —does not consist of images. Images are a later stage, as pure memory is actualized in the dynamics of the brain. (We'll examine the meaning of this.) As noted in Chapter 4, one will see critiques of Bergson and his "image memory" (e.g., Sartre, Piaget). But there is no "image memory." This is an elementary misunderstanding of the virtual (Bergson's pure memory). Image memory is only a dynamic process as the virtual actualizes via the structures of the brain.

So, we have the virtual (pure memory, which is not an image or images) extended in time. This memory is active; it is our active being. It is "driving into," affecting the brain (like water behind a dam, or in Bergson's metaphor, water behind a valve), ultimately bringing about nascent or real action, and only in this process becoming an image. This would mean:

- The virtual is bringing about a resonant, reconstructive wave-like state of the brain.
- It is specific to a past state (event) in the field.
- The field itself is permeated by this dimension of past.

We must bear in mind that perception is always, already a specification of the past, that is, specific to a past extent of the transformation of the field. Now we would have once-virtual events mixed in, driven by the active 4-D, pure memory that is equally our being. Thus, we have the auditory image, called back by memory, "Setting in motion the [elemental articulations comprising the motor diagram], and the recollection changing gradually into perception." This would be, as just noted, a gradual specification of past experienced images (here, separate words) within the 4-D field. Then, as he discussed the complexity of the recall operations: "There can be other memories realized as memory images in more or less distant centers" ([Bergson1896], p. 167).

Bergson cannot actually mean this latter statement in the way it comes across. Images cannot be in the brain. This is the essence of the whole "hard problem." There is no "photograph" in the brain. Again, we have to be conceiving of a form of specification of the past events [inwardly] "permeating" the 4-D, time-extended, holographic field.

What is the inner temporal lobe (the mental ear) receiving? As we are dealing with something (an organ) symmetric to the auditory, something which responds to the [virtual] image of words, something prior yet would have to be receiving the idea—the [virtual] image of coffee stirring. What is breaking (expressing) this idea/event into the words (The, man, stirred, the, coffee, with, spoon)? The motor diagram is the "magnet," giving the outline for this breaking and filling. The words themselves, as integral parts of the experiences of stirring (stirring, man, spoon) would be part of the image being received—the words as also an invariance across many events. Some (syntactic) mechanisms would have to pull them out, toward the diagram of sounds.

The Progression—The Virtual to Perception

In visual perception, there is a path/chain to V1, for the retina sends signals to V1. (Equally so with respect to the ear in auditory perception). So, in Bergson's speculation, there is a chain of mechanisms to the inner keyboards: (inner-V1, and (a section of?) the inner-temporal lobe), keeping in mind that by "inner" is meant some aspect of the neural structure that is oriented to the responding to the virtual. V1 is passing signals to V2, V3, V4, the motor areas, and so on, and gets a wide group of mechanisms going. So too, from the inner-V1? Yes, in Bergson's view.

We see this implication in Chomsky's syntactic unfolding: If "S" is actually a dynamic event/idea concerning stirring the coffee, the initial transition to the first rewrite rule is mysterious (Figure 7.14). The syntax expresses a learned way to break up or parse a dynamic event into phrases, words, order. Some mechanisms must be there to receive the idea (the "contemporal idea" per Lashley) and carry this out—whether producing the sentence or simply hearing and comprehending it.

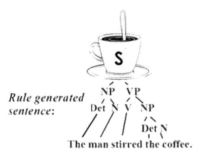

Rule generated sentence:

The man stirred the coffee.

FIGURE 7.14. Syntactic unfolding from an idea.

As we saw, the syntax unfoldment is driven by the dynamic scheme. The learning to produce the sounds of the language: also dynamic schemes, like putting together the elementary motions, in correct order, that comprise a dance step, this latter now being the dance scheme. Similarly, there are schemes for French words, English words, and so on. We know how to form syllables—consonants, vowels—for English words. These "gestures" are a somewhat different for French.

Finally, as earlier discussed, it is the whole mind—the entirety of 4-D being—that is behind the understanding of a sentence. When a speaker says to us: "The man paddled the canoe," the entirety of experience is being brought to bear. And mentioned earlier: Under the influence of LSD, this sentence would bring reverberation after reverberation of meaning as it plumbed the depths of our experience. Yes, LSD has been argued recently to be underlying the recruitment and communication of many more regions of the brain than normal [Carhart2014], and perhaps equally leads to an opening to vastly greater reception of /to past experience.

Clearly, the more we try to think this out, the more obvious that Bergson only provided a 10,000-mile Google-Earth view. Yet, it is the only theoretical attempt at a model of language comprehension, given:

- a model of perception with an explanation of the image of the external world
- where this image—our experience—cannot be stored in the brain
- where words are not stored in the brain

- perhaps even a language model that also respects the phenomena of the aphasias more deeply than most (to our knowledge, this is the case)

But as we've glimpsed, the complexity is enormous.

A *PSYCHICAL* CAUSE

An aspect of all this is the vision of a *psychical* cause—the *intent*—acting upon the matter of the brain, perhaps actualizing it as an image, setting the machinery in motion to speak a sentence or to reach for the coffee cup on the kitchen table. The "psychical cause," however, is a conceptual partition forced by the classic metaphysic of space and time. One can suspect this is why Bergson said of this, "It's a psychical cause," phraseology, "This is merely *a convenient way of expressing ourselves.*" When this classic metaphysic is analyzed, looked at, registered consciously for once, we have seen that its definition of "matter" is driven to an instantaneity of the time-extent of a mathematical point. As Bergson stated, that a mathematical point can even touch another mathematical point is nonsensical; such a point cannot cause anything. But this is the brain, as matter—a series of 3-D "objects" (like the Cubes), each with an instantaneous extent in time, acting (but since "frozen," actually incapable of doing so) one on the next. This is to say that all the machinery of the brain is in reality, rather, taking place within the indivisible transformation or flow of the universal field. But this indivisible transformation, moments permeating moments, with its intrinsic memory, has a psychical aspect in and of itself.

We visualize the intent, or the intent (taken wrongly) even as an image (reaching out for the cup), acting on the brain as a force, driving us to reach for the coffee cup. But we are back then to the discussion of force, or $f = ma$ as physics expresses it, being only an invariance law, with any notion of an actual force being explained by physics as already confusing our psychical experience of acting with what can only be an invariance law explanation, all this itself confused with trying to assign to *force* the *cause of motion*, when in fact, all analysis ends with motion (and that indivisible) being primary.

This brings us to the point where we must leave this question, for clearly, the very reluctance to conceive of a so-called "psychical cause" (mind) acting on the brain (matter) must now be deeply re-evaluated when taken within this new metaphysic of time.

REFERENCES

[Ark2020] "Intuition Behind Self-Attention Mechanism in Transformer Networks," Ark, October 17, 2020, YouTube video. https://www.youtube.com/watch?v=g2BRIuln4uc

[Asher1965] Asher, J. J. "The strategy of total physical response: An application to learning Russian," *International Review of Applied Linguistics* (1965): 3(4), 291–300.

[Asher1966] Asher, J. J. "The learning strategy of total physical response: A review." *The Modern Language Journal* (1966): 50(2), 79–84.

[Asher1967] Asher, J. J., & Price, B. "The learning strategy of total physical response," *Child Development* (1967): 38(4), 1219-1227.

[Asher1972] Asher, J. J. "The child's first language as a model for second language learning," *Modern Language Journal* (1972): 56(3), 133–139.

[Barsalou1993] Barsalou, L. W. "Flexibility, structure and linguistic vagary in concepts: Manifestations of a compositional system of perceptual symbols," in A Collins, S. Gathercole, M. Conway, & P. Morris (Eds.), *Theories of Memory.* Mahwah, NJ: Erlbaum, 1993.

Barsalou1999] Barsalou, L. W. "Perceptual symbol systems," *Behavioral and Brain Sciences* (1999): 22(4), 577–660.

[Bergson1896] Bergson, H. *Matter and Memory.* New York: Macmillan, 1896/1912.

[Bergson1912 Bergson, H. "Intellectual effort," in *Mind-Energy, Lectures and Essays* (pp. 186–230), 1912/1920.

[Behrman1994] Behrmann, M., Moscovitch, M., & Winocur, G. "Intact visual imagery and impaired visual perception in a patient with visual agnosia," *Journal of Experimental Psychology: Human Perception and Performance* (1994): 20(5), 1068–1087.

[Chalmers2022] Chalmers, D., "Are Large Language Models Sentient?" NYU Mind, Ethics, and Policy Program, October 17, 2022, YouTube video. https://www.youtube.com/watch?v=-BcuCmf00_Y

[Carhart2014] Carhart-Harris R. L., Leech, R., Hellyer, P., Shanahan, M., Feilding, A., Tagliazucchi, E., Chialvo, D., & Nutt, D. "The entropic brain: a theory conscious states informed by neuroimaging research with psychedelic drugs," *Frontiers in Neuroscience* (2014): 8, 1–22. https://doi.org/10.3389/fnhum.2014.00020

[Crockett1994] Crockett, L. J. *The Turing Test and the Frame Problem.* Norwood, NJ: Ablex, 1994.

[Elman2004] Elman, J. L. "An alternative view of the mental lexicon," *Trends in Cognitive Sciences* (2004): 8(7). 301–306.

[Dennett1992] Dennett, D. C., & Kinsbourne, M. "Time and the observer—the where and when of consciousness in the brain," *Behavior and Brain Science* (1992): 15(2): 183–201.

[Diehl2004] Diehl, R. L., Lotto, A. J., & Holt, L. "Speech Perception," *Annual Review of Psychology.* (2004): 55: 149–79.

[Fowler1986] Fowler, C. A. "An event approach to the study of speech perception from a direct-realist perspective," *Journal of Phonetics,* (1986): 14(1), 3–28.

[Fuster1994] Fuster, J. M. "In search of the engrammer. Response to Eichenbaum et al.," *Behavioral and Brain Sciences* (1994): 17, 476.

[George2023] George, D. "AGI Debate," December 23, 2022, YouTube video. https://www.youtube.com/watch?v=JGiLz_Jx9uI

[Gibson1966] Gibson, J. J. *The Senses Considered as Visual Systems.* Boston: Houghton-Mifflin, 1966.

[Karpathy2023] Karpathy, A. "Let's Build GPT: From Scratch, in Code, Spelled Out," January 17, 2023, YouTube video. https://www.youtube.com/watch?v=kCc8FmEb1nY

[Lieberman1967] Lieberman, A. M., Cooper, F. S., Shankweiler, D. P., & Studdert-Kennedy, M. "Perception of the speech code," *Psychological Review* (1967): 74, 431–461.

[Lieberman1983] Liberman, A. M., & Mattingly, I. G. "The motor theory of speech perception revised," *Cognition* (1985): 21, 1–36.

[Manning2021a] Manning, C. Stanford CS224N: NLP With Deep Learning | Winter 2021 | Lecture 1—"Intro & Word Vectors," Stanford Engineering, October 28, 2021, YouTube video. https://www.youtube.

com/watch?v=rmVRLeJRkl4[Manning2021b] Manning, C., "Stanford CS224N: NLP With Deep Learning | Winter 2021 | Lecture 9—"Self-Attention and Transformers," October 28, 2021, YouTube video. https://www.youtube.com/watch?v=ptuGllU5SQQ

[Mikolov2013] Mikolov, T., Yih, W., & Zweig, G. "Linguistic regularities in continuous space word representations," Proceedings of the 2013 Conference of the North American Chapter of the Association for Computational Linguistics: Human Language Technologies (pp. 746–751), Atlanta, GA, June 2013.

[Piaget1949] Piaget, J., & Inhelder, B. *The Child's Conception of Space*. New York: Norton, 1949/1956.

[Piantadosi2023] Piantadosi, S. "Modern Language Models Refute Chomsky's Approach to Language," Lingbuzz. https://lingbuzz.net/lingbuzz/007180

[Poeppel2022] Poeppel, D., & Idsardi, W. "We Don't Know How the Brain Stores Anything, Let Alone Words," *Trends in the Cognitive Sciences* (2022): 26(12), 1054–1055.

[Purves2010] Purves, D., & Lotto, B. *Why We See What We Do Redux: A Wholly Empirical Theory of Vision*. Oxford University Press, 2010.

[Radford2022] Radford, A., Kim, J. W., Xu, T., Brockman, G., McLeavey, C., & Sutskever, I. "Robust Speech Recognition via Large-Scale Weak Supervision," arXiv, 2022. https://arxiv.org/abs/2212.04356

[Robbins2008] Robbins, S. E. "Semantic redintegration: Ecological invariance." *Behavioral and Brain Sciences* (2008): 31(6), 726–727.

[Rogers2004] Rogers, T., & McClelland, J. *Semantic Cognition: A Parallel Distributed Processing Approach*. Cambridge, MA: MIT Press, 2004.

[Stat20223] "Word Embedding and Word2Vec, Clearly Explained," Stat Quest With Josh Starmer, March 13, 2023, YouTube video. https://www.youtube.com/watch?v=viZrOnJclY0&t=22s

[Svensson2021] Svensson, L. "Transformers—Part 1—Self-Attention: An Introduction," October 23, 2020, YouTube video. https://youtu.be/0SmNEp4zTpc

[Turvey2023] Turvey, M. "Michael T. Turvey—Interview and Reflection," in A. Szokolszky, C. Read, Z. Palatinus (Eds.), *Intellectual Journeys in Ecological Psychology: Interviews and Reflections from Pioneers in the Field.* New York: Routledge, 2023.

[Vaswani2017] Vaswani, A., Shazeer, N., Parmar, N., Uszkoreit, J., Jones, L., Gomez, A. N., Kaiser, Ł., & Polosukhin, I. "Attention is all you need," in I. Guyon, U. von Luxburg, S. Bengio, H. Wallach, R. Fergus, S. Vishwanathan, & R. Garnett (Eds.), *Advances in Neural Information Processing Systems* (pp. 5998–6008), Conference on Neural Information Processing Systems, Long Beach, CA, 2017.

[Yud2023] Yudkowsky, E. "Is Consciousness Trapped Inside GPT-4? | Lex Fridman Podcast Clips, 3," April 3, 2023, YouTube video. https://youtu.be/qplLNwboQM0

[Zhang2020] Zhang, Y., Warstadt, A., Li, H. & Bowman, R. S. "When Do You Need Billions of Words of Pretraining Data?" arXiv, 2020. https://arxiv.org/abs/2011.04946v1

CHAPTER 8

THE PROBLEM OF AFFECT

SUBJECT AND OBJECT: BERGSON'S UNIQUE PANPSYCHISM

The relation of subject and object is an epistemological question that has hardly concerned the architects of AI, and one can say this is rightfully so, rightfully so that is unless you are also claiming that AI is about to achieve AGI, or that the brain is just doing things like the GPTs only inefficiently, or that GPTs now are the best and default model of human language understanding. The nature of perception, the hard problem, the origin of the image of the external world, ultimately then, ecological intelligence—these rest precisely on one's view of this subject-object question.

In Chapter 1 of *Matter and Memory*, where his holographic model of perception was laid out, Bergson made this statement:

The relation of subject and object, in their distinction and their union, must be put in terms of time, not space. ([Bergson1896], p. 31)

What is the meaning of this Zen koan-like statement? This is where the significance emerges when the brain's biochemical dynamics are understood as specifying a scale of time upon the external field— buzzing flies versus heron-like flies versus flies as stable, motionless clouds of swirling electrons.

Consider first the property-implications of the indivisibly transforming holographic field. The state of each "point" in this field influences the states of every other point in the field, and conversely, the state of each point is the reflection of the states of all other points. We can say that at a *very elementary level*, the state of such "point" reflects a certain awareness of the rest of the field, of the states of all other points, like a vast web of awareness stretched throughout the field. This is coupled with the fact that the field is transforming indivisibly, each moment permeating the next, like a melody, where these moments are not falling into nonexistence. Looked at equivalently, the motion of the field carries an elementary property of memory, and in fact, without this elementary memory, the elementary awareness is not possible, So, we have an elementary awareness—again, *elementary*— and a "primary" memory, both as properties defined over the field.

Now we imagine our body and that of a fly within this field, taken at this elementary level, at what we can term the "null scale" of time, that is, at the smallest, most minute scale of time imaginable, even in this minute interval there yet being *change,* but for the sake of exposition we imagine there is essentially none, no change. At this scale, there is no spatial differentiation between the two bodies—our body, the fly. The distinct spatial outlines between the two are simply not there at this scale, these hard, precise outlines only existing at a much higher scale as practical demarcations, in perception, for the sake of our body's action upon them (as "separate" objects—a "spoon," a "bottle of baby formula").

Now we let our body/brain impose increasing scales of time upon the field, simultaneously then upon the fly, in effect continually *lowering* the energy state of our body/brain, or equivalently, raising the ratio of the surrounding environmental events to brain events. The fly begins to transform: from a cloud of whirling electrons, it becomes a crystalline, vibrating form, then it coalesces as a motionless, stable fly, then with wings imperceptibly moving, then to the heron-like fly, barely flapping its wings, and then to the "buzzing" being, wings ablur, of our normal scale of time. In this transformation, the spatial unity of the two in the field—our body and the fly—has never been broken, and in this, we see Bergson's principle: *subject is differentiating from object*, not in terms of space, but in terms of time.

Within the matrix of constant change that is the field in which we are embedded—objects moving here and there, spoons of oatmeal being pushed into and going away from one's mouth, being shuffled to various spots in the kitchen in our high-chair—the body is the invariant. Identity, as Bergson argued, settles upon the body. Over the course of development, as we saw Piaget describe, the body becomes (conceptually and experientially) an object among other objects, a force amidst other forces, a cause among other causes. If we wish to return to the original, nondifferentiated state, it appears, at least in the Zen-tradition, we need to begin intense concentration on a koan! The Zen master, Bassui, prescribed this one: "Who is it that sees?"

In this model, then, rather Zen-like, no one is seeing. And there is no homunculus looking at an image; we must remove the observer's eye observing the specified wave front (as in Chapter 3, Figure 3.2). Given the properties defined across the field—the elementary memory/awareness—the brain's specification of the fly as an image is simultaneously to a time-scaled, perspective-based, action-relevant *specific form of this elementary awareness*—a buzzing fly in its kitchen-environment at a scale of time.

When we speak of an elementary awareness defined across the field, taken at the null scale of time, we are not speaking yet of the experiential awareness of humans, or of frogs, or even of beetles. We are not talking here of tiny "proto-conscious" particles; we are not subject to Chalmers' "combination problem" wherein we ask how all these mini-conscious particles could possibly "combine" to form our conscious experience of the coffee cup, stirring spoon and fly buzzing by [Chalmers2016]. Whether human, frog, or chipmunk, we are dealing with the body/brain as a reconstructive wave specific to some source within the field at a scale of time, now as an *image*, that is, an aspect of the field.

This dynamical, reconstructive wave mechanism is what does away with the combination problem. It is what differentiates Bergson's panpsychism from all others of these panpsychic variants (Cf. [Skrbina2017]), for these variants have no facility, no mechanism whatsoever to explain the origin of the image (our experience) of the external world, especially at a specific scale of time. Even if one accepts the "egregious form of panpsychism" that Bishop very effectively argues one would be forced to embrace (per him, to "bite the bullet" of panpsychism) if one holds that, "any AI system is phenomenally conscious as it executes a specific set of state transitions over a finite time period" [Bishop2021], still, even apart from the blindness of this position as to the indivisible nature of flowing time, there is nothing whatsoever, even in this egregious form of panpsychism, that accounts for the origin of the image of the external world, that is, again, the origin of sentience.

In this Bergson-Gibson framework, the environment and organism (E-O) are an unbounded system, flowing within an indivisible transformation over time. This characterization of the ecological relation of organism-environment makes all the difference when considering humans versus machines.

AFFECT: WHY IT FEELS LIKE SOMETHING?

We can now look at the source of affect (emotion, pain, feelings). First though, there is something to note: We have insisted here that the more general statement of the hard problem is this: accounting for the origin of the image of the external world. The external image—filled with dynamic forms changing over time, vibrating colors at a scale of time—can certainly be taken as "something it is like." We have seen that solving the origin of the image requires an incredibly brilliant solution, one that could only come (especially *before* holography) from the mind of a being with the status of a Bergson.

This problem statement—the origin of the image of the external world—jars the mindset of philosophers. "No, no, no!" many will say, "The problem is this: explaining 'how it *feels like* something.'" Of course, this stems from Chalmers (yes, and Nagel [Nagel1974] and his "What is it like to be a bat?"):

> Why is it that when our cognitive systems engage in visual or auditory information processing, we have visual or auditory experience: the quality of deep blue, the sensation of middle C? How can we explain *why there is something it is like* to entertain a mental image, or to experience an emotion? ([Chalmers1995], p. 201, emphasis added)

And he asks: "Why doesn't all this information processing go on 'in the dark,' free of any inner feel?" (p. 203). A solution to the origin of the image of the external world—this does not even register as a problem, let alone a solution, on the mindset that has set in via this "why it feels like something" formulation.

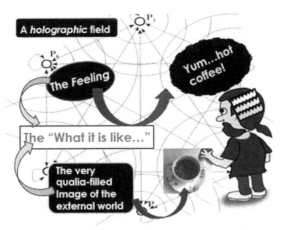

FIGURE 8.1. The symmetric halves of the hard problem.

So, yes, there is an affective component in viewing the coffee:

- Oh wow, do I need some coffee to wake up."
- "Yum, nothing like the aroma of coffee."
- "Warm up hands on the hot cup."

Our perception is permeated with these feelings. Yes, these feelings are a part of the "what it is like." But only within a framework that solves the origin of the image of the external world does the nature of affect, pain, emotion, fit, take its place (i.e., fall out naturally). These two parts are symmetric halves of the same problem (Figure 8.1). But they cannot be conflated, a conflation we will witness a bit later.

THE BERGSONIAN FRAMEWORK FOR AFFECT

We could spend some time with Bergson's 1889 discussion of affect in *Time and Free Will*, a discussion which already keys on the indivisibility of time's flow, but we will move directly to his later thought, a thought incorporating his later holographic insight into the origin of the image of the external world. In *Matter and Memory* (henceforth M&M), Bergson devotes only two paragraphs to the question of affect (in "Summary and Conclusions," Section IV, at the start). Why this little? Precisely because establishing the solution to the origin of the image of the external world and its implications for memory (where experience is not stored in the brain) is far more difficult, and truly the higher priority problem. And as just noted, affect, then, falls out naturally! Let's look at the M&M paragraphs:

> But this theory of "pure perception" had to be both qualified and completed regarding two points. For the so-called "pure" perception, which is like a fragment of reality, detached just as it is, would belong to a being unable to mingle with the perception of other bodies that of its own body, that is to say, its affections; nor would it be able to mingle with its intuition of the actual moment that of other moments, that is to say, its memory. (p. 233)

The paragraph continues but we'll stop here for a second. "Pure perception" is the holographic field taken at a single instant (not over a span of instants)—the most minute interval of time, that is, over what we just termed the "null scale" of time at the start of this chapter's discussion on Bergson's unique panpsychism, where the state of each point reflects the states of all other points. Thus, there is an elementary form of awareness (perception) defined over the field—at this null scale. There is thus far no continuity, no flow, no memory! There is just a state, then another state, then another state … . So, we have a being—but with no memory! So, continuing:

> In other words, we must begin with, and for the convenience of study, treat the living body as a mathematical point in space and conscious perception as a mathematical instant in time. (p. 233)

That is, we are yet in the framework of the classic metaphysic: Space—the continuum of points/positions, and Time as the 4th dimension of the Abstract Space—a series of instants. And:

> We then had to restore to the body its *extensity* and to perception its *duration*. By this we restored to consciousness its two subjective elements, *affectivity* and memory. (p. 233, emphasis added)

So, we've now, (a) taken the body from only being a point in the abstract space and given it a *concrete extensity* (not just a *juxtaposition* of a set of spatial points in the abstract continuum of points), and (b) set the holographic field in motion, transforming indivisibly, like a melody, with each moment (note) permeating the next and reflecting the entire preceding series. So, a primary memory is defined across the motion of the field. The body, extended, is one with this field, with no true spatial separation. Thus the (extended) *body also* has this primary memory, intrinsically. This is now the framework of the temporal metaphysic. Why does he say "affectivity" was restored? In the next paragraph he begins with this: "What is an affection? Our perception, we said, indicates the possible action of our body on others" (p. 233).

Here we need recall Bergson's description of the holographic field as the sea of real actions, where each object is obliged:

> ...to transmit the whole of what it receives, to oppose every action with an equal and contrary reaction, to be, in short, merely the road by which pass, in every direction the modifications, or what can be termed real actions propagated throughout the immensity of the entire universe. (p. 28)

In this sea of motion, the body is selecting those actions of interest to itself, that on which it can act—this as the specified image, an image equally being virtual action relative to the specified scale of time. He goes on:

> But our body, being extended, is capable of acting upon itself as well as upon other bodies. Into our perception, then, something of our body must enter. When we are dealing with external bodies, these are, by hypothesis, separated from ours by a space, greater or lesser, which measures the remoteness in time of their promise or menace: this is why our perception of these bodies indicates only possible [virtual] actions. (p. 233)

In Figure 8.2, we visualize a "remote body" (the wasp) separate from our body but closing in by less and less time and distance (until the sting). Bergson continues:

But the more the distance diminishes between these bodies and our own, the more the possible action tends to transform itself into a *real action*, and the call for action becoming more urgent in the measure and proportion the distance diminishes. And when this distance is nil, that is to say, when the body to be perceived is our own body, *it is a real and no longer a virtual action* that our perception sketches out. (p. 233, emphasis added)

FIGURE 8.2. Affect as *real* action.

This real action is the sting/OW! part of Figure 8.2. The wasp is still external (excepting his stinger), still virtual action; the pain within the arm from the sting is internal. The body is *integrally part* of the sea of real actions. These are melodically, indivisibly transforming – *not* instant after discrete instant, nor computing "state" after "state." This is the basis of the body's affect!

> Such is, precisely, the nature of pain, an actual effort of the damaged part to set things to rights, an effort that is local, isolated, and condemned to failure in *an organism that can no longer act except as a whole.* (p. 233, emphasis added)

This is *not* a machine, not a device consisting of independent parts that have been designed to work together. This organism is an *undivided whole*.

> Pain is therefore in the place where it is felt, as the object is at the place where it is perceived. Between the affection

> felt and the image perceived there is this difference, that
> the affection is within our body, the image outside our body.
> And that is why *the surface of our body*, the common limit
> of this and other bodies, *is given to us in the form both of
> sensations and of an image*. (p. 234, emphasis added)

This is Bergson's framework for affect. At this point, we are going
to place Bergson within a recent work by Solms [Solms2021], who,
relying heavily on Friston's framework, has provided us with a very
recent theory on affect.

SOLMS ON AFFECT

Solms' framework is that of *Predictive Processing*, and he is an
advocate of Karl Friston's approach to this [Friston2010]. In
general, the brain-system is seen as generating predictions, both of
the external world (objects/events) and of its internal world (affects).
These are continually matched to the input received, and the error
(i.e., the discrepancy between prediction and input) is checked.
Input is *suppressed* (in favor of the prediction). Thus, for perception,
if there is a match, we are literally only seeing the internal memory
(predicted) image, else, the projected memory image is "adjusted."

Where do these memory images come from? There is *no* initial
perceptual image to be "stored." One will search Solms in vain for an
answer. But moving on, Solms notes that a heavily favored concept
(late 1800s) was that the *cortex* is the seat of consciousness. Early
theorists, he states, insisted that experience consisted of memory
images produced by the cortex. "It is important to recognize that
the mind, on this view, consists entirely of memory images that
reflect past experience of the outside world", and "Incoming
sensations merely stimulate these images and their associations into
consciousness" (p. 60). (Again, one wonders, how did they not ask
where the images to be stored in memory first came from?)

But Solms starts early in the book with vignettes of
hydraencephalic children. These are born with virtually no cortex.
Yet, they are obviously conscious, have feelings, joy; they love what
they are doing. So, the cortex cannot be entirely critical.

FEELINGS

Solms states that, "When we look at ... our own consciousness ... there are general categories of content. There are, of course, 'representations' of the outside world—both perceptions and memories of it and thoughts about it ..." (p. 88). Philosophers have heavily attended to these, but Solms emphasizes that there are also feelings about what is going on in the world, feelings reflecting the condition of our bodies, not to mention emotions and moods that qualify our experience of the world and shape our behavior within it.

He notes the actual importance of this: "This is what affects are for: they convey which biological things are going well or badly for us, and they arouse us to do something about them" (pp. 98–99). The *arousal* is the key point here; it is why affective sensations are different than perceptual ones. The second key, per Solms, is that feelings are always conscious, to the point that an unconscious feeling is not a feeling.

There is a third thing about feelings, for there is a prioritization: felt needs are prioritized over unfelt ones; a selection must be made, and this selection is determined by the relative strengths of one's needs, and the size of the error signals in relation to the range of opportunities afforded by your current circumstances. The prioritization of needs, when realized, leads to governing your voluntary behavior. "You decide what to do and what not to do based on the felt consequences of your actions ... [this conscious choice] reveals the deepest biological function of feeling: it guides our behavior ..." (p. 100).

THE SOURCE OF FEELING

There is a feature of the brain's physiology that makes this true. For Solms this is in the brain stem. Thus, he points to positron emission tomography during states of GRIEF, SEEKING, RAGE, FEAR (he capitalizes to show their fundamental nature), which show that the highest metabolic activity occurs in the core brain stem (and other subcortical regions). There are also the lesion studies, deep brain stimulation, pharmacological manipulation, and functional neuroimaging. These "all point to the same conclusion: the reticulate core of the brainstem generates affect" (p. 124).

This part of the brain, the reticulate core, Solms notes, is the only part we know of that is necessary for arousing consciousness as a whole. It has an equally powerful influence on feeling and these feelings pervade conscious experience. In fact, Solms views our dealing with feelings (which come from within and regulate our biological needs) as one of the central tasks of consciousness, where, as would be entirely consonant with Bergson's difference between affect and vision as being simply real versus virtual action, the neurological sources of affect and consciousness are deeply entangled.

Solms asks: "What should we call this basic medium, this mysterious mind-stuff that seems to well up within us?" (p. 125). He settles on the term "arousal." This he relates to various cortical rhythms: alpha, delta, theta, beta, gamma. He states: "The cortex becomes conscious only to the extent that it is aroused by the brainstem" (p. 128).

THE CONFLATION WITH THE HARD PROBLEM

Current circumstances, Solms notes, must be taken into account. This is effected in the nearby superior colliculi. Here is held a multi-sensory map—a "massively compressed and integrated representation of the exteroceptive world." This is the objective, *external* state. Here too is a map that controls eye movements. It is intrinsically more stable, and other maps are calibrated relative to it. (He holds this is why the world is stable while the eyes saccade.) And we hit the inevitable indirect realist position: "The stabilized scene also hints at the fact that we perceive just that—a scene—*a constructed perspective of reality, not reality itself*" (p. 140, emphasis added). This goes along with: "The brain's internal model is the map we use to navigate the world—indeed *to generate an expected world*" (p. 144, emphasis added).

So, the problem of the origin of the image of the external world is essentially being equivalenced to the origin of feeling. Both are being *generated* by the brain.

ADDING FRISTON AND FREE ENERGY

Solms will spend many more chapters (Chapters 7–12) on explaining this solution to the hard problem. However, the model is not going to get any better than this. This is all integrated within Friston's [Friston2010] Predictive Processing framework (henceforth, PP) and his Free Energy Principle (FEP). We want to take a brief look at this, for Friston and his FEP is quite active in AI discussions and the PP framework is currently considered a leading candidate to explain mind.

Friston's framework starts with homeostasis. Living organisms must occupy a limited range of states: we cannot get too cold, or too hot, so, a limited range of temperatures. This brings in entropy, for homeostasis is going against entropy, that is, against the system's occupation of all possible states. This in turn brings in the "Markov Blanket," wherein we have four states: internal, active, sensory, external (Figure 8.3). Sensory states feedback the effect of "active" states on the external states, this to adjust further action, or, put more concretely, this "feedback" statement is aimed at capturing the perception-action cycle wherein our action changes something in the external world (like the coffee surface via the spoon) and we "adjust" (because the liquid is splashing).

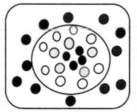

A system of concentrically layered structures.
Each with an inner core and an outer surface.

- The states of the white layer (white circles) are influenced by the outer (black), and white influences the gray, but gray does not influence the (outer) black.
- The states of the gray layer are influenced by the inner core (black), and influence the white.
  - The external states (black) can only be "sensed" vicariously by internal ones as states of the blanket.

FIGURE 8.3. Friston's Markov blanket schema (adapted from Solms, 2021).

There is no mechanism for perception in this Markov blanket framework, other than perhaps, some hope for "emergence." In other words, the system is blind. The external world must be *inferred* (and *somehow* the image of the coffee cup and stirring spoon come to exist *as an inference*). Helmholtz, a prime exponent of inference, would certainly approve. This brings in the Bayesian inference apparatus:

- predictions on the external world (with some prior probability)
- error comparison—prediction against actual signal
- adjustment of prediction

Thus, we come to Friston's "free energy principle" (Figure 8.4). In this framework, the system always aims to minimize F (free energy). Now (formally) add in needs and affects driving/weighting the predictions, and we have pretty much covered it, albeit "it" being a very vast, complex structure. And yes, given the equivalencing of F to the "upper bound on surprise" (re predictions), then conversely, when there *is* surprise, this sounds very much like Peirce and his start of the operation of abduction, but sans any registration of events (in the FEP framework) as invariance structures, or "stacks" of events in 4-D experience, or a mechanism for redintegration, or a basis for a concrete, felt dissonance underlying "surprise."

The Bayesian framework for F.

$$F \equiv \int dT \left[q(T) \ln \frac{q(T)}{P(T,S)} \right]$$

F = Free Energy

Here, T is temperature.
q(T) = our *guess* about the Temp.
S = the actual environmental state.

- F is the upper bound on "surprise."
  Or, F is the *surprisal* we experience on sampling data, given a generative model.

- F is *analogized* to Helmholtz's physical, thermodynamics (for a system);.
  F = total amount of energy – energy already employed in effective work.

Recast:
  - F = Avg. energy – entropy
- Avg. Energy = expected probability of an event, and entropy = *actual* incidence of event.

FIGURE 8.4. Scheme of the free energy principle. (Adapted from Solms, 2021.)

This FEP-structure acts as a (static) state to (static) state *comparator*. In this comparator process (Figure 8.5) we have the following steps:

1. Project (*generate*) the prediction as an image. (HOW? How does the brain generate an image from its neural structure and project it to the outside world? This is *magic*.

2. *Compare* the prediction against the actual external environment event, get the degree of error.

3. Re-project a new prediction.

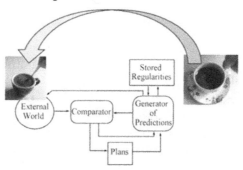

- Take snapshot of event, compare
- Project image of next state
- Meanwhile the event is/was ongoing!
- And...you only "see" the "projected" (the prediction) static state.

FIGURE 8.5. The event comparator process.

How would this ever work for an actual, physical, dynamically changing event? Essentially, the preceding steps come down to this:

- Predict (*generate*) the image of the event (stirring the coffee).
- Take a snapshot (static) of the actual event (stirring the coffee).
- Compare the two (find the degree of error), adjust the image.
- Meanwhile the event is ongoing!
- Project (*generate*) the image of the next "state" of the event.
- Meanwhile the event is ongoing!
- And one only "sees" each "projected" (predicted) static state.

If Bergson had to ask if the memory theorists of his day had contemplated the nature of words and sentences, we must ask the equivalent re the nature of *events*. One must contemplate here what we are attempting to "match" (Figure 8.6). The (present) coffee-stirring event is a dynamic invariance structure, constantly transforming, with the "wielding" of the spoon being and carried by dynamic haptic flow fields, to include adiabatic invariants, inertial tensors, and more. The (past) event from which we are supposedly drawing the image as a prediction, is (and was) itself also a dynamically changing field, defined by invariants. How are the two dynamically transforming, multimodal fields ever synced up to check for a *match*? The two will *never* match, unless perhaps allowing for a massive degree of "error." We saw effectively the same problem with DORA (Chapter 4). What does "matching" actually mean? It only makes sense within a static and highly make-believe, highly artificial framework of thought.

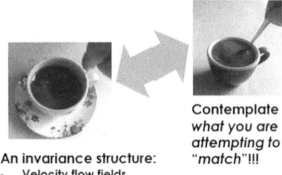

Contemplate
*what you are
attempting to
"match"*!!!

An invariance structure:

- Velocity flow fields
- Adiabatic ratio (E of oscillation/frequency)
- Inertial tensor (various momenta)
- Acoustical invariants
- Texture gradients/ratios/flows

FIGURE 8.6. Matching two dynamic events?

This predictive process is taking samples and snapshots of an event. We can ask, what if the cup were our cubical, strobed out-of-phase, rotating cup (Chapter 2)? Again, form, or the event as form, for the brain, does not exist in an instantaneous sample.

THE HARD PROBLEM AND FEELINGS

So, just neglecting any further the embedding in Friston's framework, in Solms' notion of affect—its brain mechanisms, the activation/ arousal of the cortex—we see little that is not commensurate with Bergson and his short comments on affect, with the exception of the conflation noted. Again, biophysical dynamics equals *real action* within the body. But Solms is trying to explain both: (1) affect, and (2) the origin of the image of the external world via one mechanism—neural generation. Discussing Chalmers (in Solms, Chapter 10) Solms simply accepts the Chalmers formulation: "Why should physical processing give rise to a rich inner life at all?" (p. 239). But the way the question is posed, namely, *physical* processing (as "physical" is currently understood), already presupposes the route (or lack thereof) to the solution, and exactly the route Solms/Friston took.

Bergson essentially redefined the "physical" in terms of:

- Extensity—Undivided, *not* the abstract, infinitely divisible continuum of points
- Duration—an indivisible transformation of the holographic field (*not* instants)

This allowed conceiving of the brain as acting as a reconstructive wave, passing through this always indivisibly transforming field, specifying a past portion of the field (as an image) at a scale of time. And yes, we can modify this just previous statement per Solms: "the brain, *when appropriately aroused*, acting as a reconstructive wave…" But Solms and Friston reject this path at the start: "We must go beyond the disciplinary constraints of psychology and physiology. We scientists typically do so by turning not to metaphysics but to physics" (p. 147). And if your metaphysics is precisely the problem?

ENGINEERING CONSCIOUSNESS, ENTROPY AND MATHEMATICAL LAW

Solms and Friston turned to physics, as did Hameroff [Hameroff1996], McFadden [McFadden2002], and others to solve the hard problem. But … they did not really turn to physics! (Yes, they bring in entropy, free energy, homeostasis …) Rather, they turned to mathematics; they built a mathematical system as a blueprint for the brain.

There is a huge difference between:

1. Creating a mathematical system, then using *any* physical components available for constructing the system so long as the logical operations are implemented. This is the Turing Machine: a mathematical system that can be supported with any components—as long as the logical operations are carried out, the latter being the only truly relevant operations.

2. Describing a concrete system with mathematics, as best one can, for precision in building with real components, forces, metals, chemistry. Example: Steinmetz (*Theory and Calculation of Alternating Current Phenomena*, 1900) formulating the laws for building Tesla's AC motor (very hard to recreate at the time without it overheating and blowing up).

In item 1, the math is the "reality": Build a physical system that does the math operations. In item 2, the physical system is the reality. Mathematics at best (partially) describes it.

So, the Solms/Friston intent, along with many a worker in AI, is to *engineer* consciousness: "...in principle, an artificially conscious self-evidencing system can be engineered. Consciousness can be produced. This will realize the wildest dreams of Helmholtz" ([Solms2021], p. 305).

Surely, the FEP with its reliance on entropy will be the key building block. This brings us back to our discussion of the unreality of mathematical law (in Chapter 7). For Bergson, we saw mathematical law expresses no *positive* reality: "It is merely this negative tendency that the particular laws of the physical world express. None of them, taken separately, has objective reality" ([Bergson1907], p. 218).

He noted that each law is the work of an investigator who:

- isolated certain variables
- worked from a certain bias
- applied conventional units of measurement

And we saw him noting, "measuring is a wholly human operation, which implies that really or ideally we superpose two objects one on another a certain number of times" (p. 218). As he said, this is a superpostion of which "nature did not dream." Thus, physics is relative to (a) the variables it has chosen, and (b) the order of the problems examined. Put differently, these laws are *anthropomorphic;* they have no *ontological* status.

So, is there a "law of entropy" that Friston and Solms can claim is available to the brain which will underlie its processing and the crucial FEP that can be used to engineer the brain? Or, bringing in another theorist and his simulation theory (in his book, *Reality+* [Chalmers2022]), is there a set of mathematical laws, to include a law of entropy, that Chalmers' cosmic simulator (located somewhere, no one knows where) will employ as its mathematical guide for generating (constructing) the universe? After all, we would not want the coffee cup, once laying shattered in pieces on the kitchen table, or the egg broken and splattered on the floor, to constantly be reassembling, neither in Chalmers' simulated world nor in Friston's perceptual world. Is there such a law? Not really. This was the subject of a famous paper by Jaynes ("Gibbs vs. Boltzmann Entropies" [Jaynes1963]). Jaynes noted initially that, "...it may come as a shock to realize that nevertheless thermodynamics knows of no such concept as 'the entropy of a physical system.' ... a given physical system corresponds to many thermodynamic systems" ([Jaynes1963], p. 391).

Jaynes has us imagine a crystal of Rochelle salt. There can be multiple experiments upon it. In one set of experiments, temperature, pressure, and volume are key, and the entropy can be expressed as some function of $S_e(T, P)$. In another set of experiments, the keys are temperature, the component e_{xy} of the strain tensor, and the component P_z of electric polarization. Now the entropy is a function $S_0(T, e_{xy}, P_z)$. "It is clearly meaningless to ask, 'What is the entropy of the crystal?' unless we first specify the set of parameters which define its thermodynamic state" ([Jaynes1963], p. 391).

Jaynes acknowledges the argument that these experiments only use *some* of the degrees of freedom available to a system, there being a "true" entropy which is defined across *all* these possible parameters simultaneously. He destroys this argument, observing that we can always introduce more degrees of freedom. We can expand each element of the strain tensor in a complete orthogonal set of functions $\varphi_k(x, y.z)$:

$$e_{ij}(x.y.z) = \sum ka_{ijk}\varphi k(x, y, z)$$

Now, we can vary each of the first 1,000 coefficients a_{ijk} independently, the crystal being now a thermodynamic system of over 1,000 degrees of freedom. Still, he notes, we nevertheless believe the laws of thermodynamics would hold. So, the entropy must be a function of over 1,000 independent variables. Jaynes sees, *"no end to this search for the ultimate, 'true' entropy until we have reached the point where we control the location of each atom independently"* (p. 392). But at this point, wherein we control every last atom, "the notion of entropy collapses, and we are no longer talking thermodynamics" (p. 392).

Jaynes goes on, echoing Bergson, for from all this, we see that entropy is an anthropomorphic concept. This is not only from the statistical concept that it measures the extent of human ignorance as to the microstate, but in this: *"Even at the purely phenomenological level, entropy is an anthropomorphic concept.* For it is a property, not of the physical system, but of the particular experiments you or I choose to perform on it" (p.393).[1]

So, whether we really own that set of mathematical laws from which a Great Simulator can generate a physical universe (fooling us all) or from which, just as difficult, a Karl Friston can fashion a brain—this, we would hold, is a very questionable position.

APPROXIMATION IS NOT ENOUGH

Underlying the Friston-Solms vision is this: "Consciousness is part of nature, and it is mathematically tractable" ([Solms2021], p. 304). But nature transforms indivisibly. It is not mathematically tractable; only in approximation, only in the *ideal limit* that the abstract space of the classic metaphysic represents. *And approximation is not enough.* It is this melodic, indivisible motion, where each "instant" permeates the next, where each "instant" builds from the preceding that characterizes the *real motion* of the field, and the body is embedded in this. And this underlies affect! Without this, there is none. And none (zero affect) is exactly what Friston and Solms will have.

Take the quality of "mellow," a quality which is also a *feeling*.

- The violin is mellow. (Sound)
- The room is mellow. (Visual)

---

[1] So, yes, although Bergson saw "no artifice of measurement" in the concept of entropy *as he stated it,* in taking the law to a more precise, experimental form, Jaynes shows entropy to be just as anthropomorphic as the other laws of physics of which Bergson was thinking.

- The wine is mellow. (Taste)
- The kitty cat is mellow. (A personality, a mode of being)

Mellowness can only build in a melodic, building, indivisible time. Each note (or "state") of a "mellow" song reflects the history of the song, of all the previous notes of the song. Hold one note in "Twinkle, Twinkle Litte Star" for an extra amount of time: the quality of the phrase changes. Mellow has nothing to do with being an abstract node in a connectionist net, pointing, via weights, to rooms, wines, violins, kitty cats. This is an invariant, call it a "qualitative invariant," defined over multiple modalities, each transforming in an indivisible—"instant" permeating "instant" —time. The classic metaphysic and its mathematics cannot support this, that is, it cannot support affect! It cannot *approximate* it.

Physics with its mathematics is not the safe haven that Solms and Friston believe, at least not in its current form, where mathematics is considered *explanation*. Tracing the history of this conception through mathematics and physics is too much for here, but it is something AI should be aware of. We intimated this in Chapter 1; Newton had expressed his position that his gravitational equation, $F = G(m_1 m_2)/r^2$, was *not* a *physical* explanation of gravity's force, but rather a *description*, saying, "Hypotheses non fingo." But one can trace the furtherance of the mathematization through Euler-Lagrange, where the Lagrangian was taken as an *explanation* of the path that a falling object takes through a gravitational field, rather than, again, only a *description*. It continued with Einstein's simply *geometrizing* the same equation (cf. Carroll's discussion [Carroll2021], also [Robbins2022]), none of which would have impressed Newton as a *physical* explanation; he *still* would say of gravity, "I feign no hypotheses." At the heart of this mathematics is *approximation*. As earlier noted, even the wave equation, ψ, supposedly at the heart of reality, relies on this approximation, namely, on Euler's identity, $e^{i\pi} = -1$, and the limit-taking required therein for that wave-cycle to return to its starting point, unchanged.[2]

[2.] The "it from bit" conception [Wheeler1991] is a correlated problem. The "bit" is entirely an *abstraction,* having a "reality" only via very concrete physical implementations, for example, magnetic cores and wires, or complex breadboards with transistors, or, for qubits, LC circuits—and *interpreted* as supporting such (that is, "bits" or "qubits").

We have taken the abstract space, in essence only an ideal limit, as the start—the foundation—of reality. But it is not; it is a *derivative* of our cognitive development. It is a derivative Piaget tried to chronicle and document. It cannot be the start of the theory of consciousness, of feeling, of experience.

REFERENCES

[Barbour2010?] Barbour, J., "Bit from It," bit_from_it.pdf (platonia.com), 2010? (Note: The actual year when written is hard to determine.)

[Bergson1896] Bergson, H. *Matter and Memory*. New York: Macmillan, 1896/1912.

[Bergson1907] Bergson, H. *Creative Evolution*. New York: Holt, 1907/1911.

[Bishop2021] Bishop, M. "Artificial intelligence is stupid and causal reasoning will not fix it," *Frontiers in Psychology* (2021): *11*, 1–38.

[Carroll2022] Carroll, S. *The Biggest Ideas in the Universe: Space, Time and Motion*. New York: Penguin Random House, 2022.

[Chalmers1995] Chalmers, D. "Facing up to the problem of consciousness," *Journal of Consciousness Studies* (1995): 2(3), 200–219.

[Chalmers2016] Chalmers, D., "The combination problem for Panpsychism," in G. Brüntrup, & L. Jaskolla (Eds.), *Panpsychism: Contemporary Perspectives*. Oxford, England: Oxford University Press, 2016.

[Chalmers2022] Chalmers, D. *Reality+: Virtual Worlds and the Problems of Philosophy*. New York: W. W. Norton and Company, 2022.

[Friston2010] Friston, K. "The free-energy principle: a unified brain theory?" *Nature Reviews Neuroscience* (2010): *11*(2), 127–138.

[Hameroff1996] Hameroff, S., and Penrose, R. "Conscious events as orchestrated space-time selections," *Journal of Consciousness Studies* (1996): 3(1), 36–53.

Barbour [Barbour2010?], for one, has come to see this. That "bits" can exist on the massively destructive surface of a black hole – this is a pure mathematical fiction (Cf. [Robbins2019]), as though the "bit" abstraction had, using Bergson's phrase, some "positive reality."

[Jaynes1963] Jaynes, E. T. "Gibbs vs Boltzmann Entropies," *American Journal of Physics* (1963): 33(3), 391–398,

[McFadden2002] McFadden, J. "Synchronous firing and its influence on the brain's electromagnetic field: Evidence for an electromagnetic field theory of consciousness," *Journal of Consciousness Studies* (2002): 9(4), 23–50.

[Nagel1974] Nagel, T.M "What is it like to be a bat?" *The Philosophical Review*, (1974): 83(4), 435–450.

[Robbins2019] Robbins, S. E., "Bergson's Holographic Theory -39 - The Holographic Principle," March 29, 2019, YouTube video. https://www.youtube.com/watch?v=PlDHuGk7B8g

[Robbins2023] Robbins, S. E. "Bergson's Holographic Theory—79—General Relativity," March 31, 2023, YouTube video. https://www.youtube.com/watch?v=xJlw4-3jyQA&t=3s

[Skrbina2017] Skrbina, D. F. *Panpsychism in the West.* Cambridge, MA: MIT Press, 2017.

[Solms2021] Solms, M. *The Hidden Spring: A Journey to the Source of Consciousness.* New York: W. W Norton and Company, 2021.

[Wheeler1991] Wheeler, J. A., "Sakharov revisited: 'It from Bit'," (Ed.) M. Man'ko, *Proceedings of the First International A D Sakharov Memorial Conference on Physics*, Moscow, USSR, 1991, May 27–31, Nova Science Publishers, Commack, NY, 1991.

CHAPTER 9

Space, Time, and the Requirements for a Conscious Device

ECOLOGICAL SCALE AND ADVANCED TURING TESTS

There have been various suggestions for improvements on the Turing Test. There are the "mini tests" of Judea Pearl, probing the understanding of causality [Pearl2018], or the Total Turing Test version of Harnad [Harnad1991]. We will just add some extra considerations here. We noted earlier that in the Bergson-Gibson framework, the environment and organism (E-O) are an *unbounded* system, flowing within an indivisible transformation over time, and that this characterization of the ecological relation of organism-environment makes all the difference when considering humans versus machines.

Test 1

Let us start with the strobed, rotating, wire-edged cube, rotating, we shall say, one complete revolution per second (Chapter 2, Figure 2.6). Strobed in phase with its symmetry period, it is perceived as a rigid, rotating cube. Strobed out of phase, it is a wobbly, plastically deforming, non-rigid, not-a-cube. This, we would argue, is the effect of direct specification within an unbounded, indivisibly transforming, E-O system. The rotating cube with its four-fold symmetry, strobed in phase, is "emitting" (so to speak) a regular, temporal pulse coordinate with the cubical form's intrinsic spatial symmetry. With the irregular strobe, the temporal symmetry, coordinate with the spatial symmetry, is being destroyed. As the specification—always probabilistic, always optimal—is to a past extent of the whole E-O system, the entire transformation is specified in this deformed way, consistent with an irregular, nonrigid form.

What would we expect if we did this same experiment on a "perceiving" robot? The robot is sampling this transformation with high speed "shutter" at some rate. The rate does not matter; let us say ten samples/second. Each sample, near instantaneous, when pixelated, becomes an image of a cube, each in a slightly different orientation. There seems no reason why the robot would not simply continue to "perceive" a cube, no matter what the periodicity of the strobe rate.

As in this E-O system, perception is virtual action, a human is equally perceiving a form of possible action on what is now a nonrigid, plastically deforming object—an interesting form of necessary modulation of the hand-arm. Again, for the robot, if asked to grasp the cube, it is doubtful it would try any such adjustment in its motion.

An interesting question arises if we take this to the rigid rotating ellipse with its perimeter of velocity of vectors. Rotating too fast, per Weiss et al. (Chapter 2), it violates their "motion is slow and smooth" constraint (applied mathematically in their computational model) and the ellipse loses its rigidity (Mussati's illusion) [Weiss2002]. As this is a computational model, applied over velocity flow fields, one could ask, would the (computational) robot not see the same thing if the speed of the rotating ellipse violated this constraint, that is, would it not also see a nonrigid ellipse? We would be inclined to say, "In what sense is it seeing?" Its computations have indicated that the ellipse should be nonrigid, and the robot could give a right answer, "I see a nonrigid ellipse." It might well, given programming, adjust its hand-arm to act relative to this "illusion." But even given the velocity flow-basis of the computation, it is still sampling the rotation, each sample being, once pixelated, an "image" (still only a matrix of bits) of a rigid ellipse, and the computation becomes but a gloss on this series, *indicating* that the ellipse is, or better, *should* be conceived as, and stated by the robot to be, nonrigid.

Can the strobed, rotating cube, like the ellipse, be given a computational model computed over velocity flows? Even so, the above problem, as we have taken it in the rotating ellipse context, still remains.

Test 2

We imagine a drone flying over a course. Wherever there are two cubes, ten feet apart from each other along the course, the drone must fly between them. We know that the drone is sampling the layout at twenty samples/per second. Each sample, processed, is pixelated as an image of a cube. Now we set the cubes rotating at twenty full revolutions per second. For a human, flying the course with a jet pack, these are now spinning cylinders with a fuzzy haze surrounding a more solid core; he does not bother to fly between them. What will the drone "perceive"?

Switching this situation slightly, we set the cubes spinning initially, turning them into perceived cylinders. The human, in his normal state (his normal scale of time), does not bother flying between them. We suspect the robot, without code kluges, is flying through them, but if not, there is still this test: We give the human a certain dosage of LSD, a dosage we know changes the perceived scale of time such that the spinning cylinders have slowed—they are now either slowly rotating cubes or even stable cubes. The human, albeit now flying over a rather globally "transformed" course environment (with maybe some buildings "melting"), now flies between the cubes. Now we would love to give the same dosage to the robot, for we would expect Mr. Robot also to now be flying between the cubes. Alas, we cannot, for the engineers of the AI/robot have no place for biochemistry.

Test 3

Elsewhere [Robbins2006], we've explained several effects in memory research in the verbal learning tradition as noted in Chapter 4, for example, the *concreteness* effect involving concrete words versus abstract words (the effect has to do with invariance structures and fine tuning of the reconstructive wave), the enhancing effect of imagery applied to these memory tasks, the excellent memory performance in the very ecological subject-performed tasks [Robbins2006]. Now imagine giving an AI, even a robot-AI, one of these tasks (there is a huge number of variants within the verbal learning tradition). Let's say the task is to "remember" a list of twenty paired-associate word-pairs consisting of ten abstract pairs, ten concrete pairs. Would we expect to see the standard human performance in this task reflected in the AI's performance? And would we expect to see the human differential performance reflected if a robot-AI *acted out* twenty pairs as in the subject-performed tasks versus being instructed to simply remember them (i.e., simply being given verbally the words SPOON-COFFEE as a pair to remember versus concretely stirring the coffee with a spoon)? Well, obviously this task is a sort of joke; the test is absurd for an AI, the AI would simply spit back the twenty pairs from its "memory." Unfortunately, we cannot simply ignore the mass of findings from human memory research and yet make the claim that AI is "doing the same thing as the human brain (only better)."

This is just a tiny fraction of the tests one could come up with if one went into the areas of perception research (e.g., how will AI reflect the plethora of illusions—visual, auditory, kinesthetic?), memory research, cognition. With no concrete dynamics underlying its perception, reflecting human performance is not possible.

HAVING TEA IN THE CHINESE ROOM

No discussion around the subject of AI appears to be complete without some attention to the Chinese Room. Searle's argument (or very clever "intuition pump" as Harnad called it [Hayes1992], p. 221) relied on one primary point, namely, an appeal to the implausibility of any understanding (of Chinese) arising (in any mysterious "space"—mental, perceptual) via Searle's mere syntactic manipulation of symbols as Searle "translated" Chinese statements to English in his room. As to what "understanding" actually might be, other than it might require, (a) consciousness (with no theory of this), and (b) common sense background knowledge (with no theory of this), the Chinese room argument had not a great deal in a positive way to offer as far as an alternative theory of mind, and this deficiency in the argument has been the open door to the voluminous controversy ever since, in fact, per the *Stanford Encyclopedia of Philosophy* discussion of the Chinese Room, "By 1991 computer scientist Pat Hayes had defined Cognitive Science as the ongoing research project of refuting Searle's argument" [StanfordRoom].

The core problem with the argument and the controversy has been the complete neglect of the problem of time. One can read the long *Stanford Encyclopedia of Philosophy* entry on the Chinese Room subject (as of 2023) and never even see the *word* time. Searle has never, to our knowledge, explicitly examined the classic metaphysic of space and time, but this framework, as we have seen, is the foundation of the computer model of mind. Yes, as Searle stated, "Syntax is not sufficient for semantics," but the core of this insufficiency is that syntax is simply rules for the concatenation or juxtaposition of "objects" (be it "marks," "bits," etc.) in an abstract space (taking a definition from [Ingerman1966]) in an abstract time. It is a purely spatial operation. Time, the concrete, indivisibly flowing time, or duration, means nothing to these operations. The machine can carry out the same 100,000 operations in a millisecond, or in an hour; it makes no difference, the validity of the algorithm is unaffected, the end result is unaffected, the computation is achieved. This is why the argument, once made against Searle, that "surely the machine would understand" because of its processing speed versus Searle's turtle-like slowness in manipulating his symbols, is absurd—a failure to grasp both the problem of time and the definition of the computing machine.

Sitting in the Chinese room, we stir the tea in the teacup in an indivisible flow. The invariants of this event are defined over the indivisible, melodic motion. We reach out and grasp the teacup in an indivisible movement. The feel of the tea flowing down the throat, its delicious taste, are via our embeddedness in the indivisible, real motion of the entire universal field with its interpenetrating, building moments. The linguistic string, "She stirred the tea, reached for the cup, and took a sip," is designed to redintegrate a form of this flowing experience in its indivisible time. This already implies a theory of experience—the origin of the ever-changing image of the external world—completely missing in Searle, in fact, in the whole debate.[1] But once this is understood, the concept that such a sentence is ever available in its actual meaning to the purely syntactic operations of a machine becomes a wonder, a wonder that such an idea was ever debated.

[1.] Searle tried to argue for and describe a theory of direct perception [Searle2015], never once mentioning Gibson, while Bergson certainly failed to make his list of "seven great philosophers of perception" in the same work, but his failure to even consider the problem of time undercut any possibility of validity for his theory or help for his "room" argument.

So, when the question is debated, "Does ChatGPT (or some GPT-[n]) actually *understand* those sentences and paragraphs and books it is putting out?" it is simply wrong to avoid the question by saying, "No one knows what 'understanding' is," and move on as though there is no real question worth considering [Shapiro2023]. Understanding, whether Penrose's proofs or Wertheimer's proofs, or, of "She stirred the tea," is the perception of an event with its invariance defined over an indivisible flow of time. Without integral embeddedness in this flow, there is no understanding.

THE REQUIREMENTS FOR A CONSCIOUS "DEVICE"

Implicitly, along the way, we have been developing a set of requirements for the creation of a conscious "device." These are the requirements—a minimum set of requirements—that must be met if we are to create an intelligent "device" equivalent to a human in intelligence. They are obviously very high-level requirements. They imply a vast amount of physics and engineering to work through, where the emphasis is on engineering—the implementation of a real, concrete dynamics—not software. This dynamics must function integrally within the indivisible flow of time. The requirements look like this:

1. The total dynamics of the system must be proportionally related to the events of the matter-field such that a scale of time is defined upon this field.

2. The dynamics of the system must be structurally related to the events of the matter-field, that is, reflective of the invariance laws defined over the time-extended events of the field.

3. The information resonant over the dynamical structure (or "states") must integrally include relation to, or feedback from, systems for the preparation of action (for from the vast information in the holographic field, the principle of selection is via relation to possible action by the body).

4. The global dynamics must support a reconstructive wave.

5. The operative dynamics of the system must be an integral part of the indivisible, nondifferentiable motion of the matter-field in which it is embedded.

6. The operative dynamics of memory retrieval must rely on the invariance structure of events as the modulating mechanism of the device.

7. The operative dynamics must support an articulated simultaneity within the non-differentiable flow of time.

8. The operative dynamics must achieve a trajectory supporting COST and the symbolic function.

9. The operative dynamics must support abstraction as a function of invariance defined over concrete experience (or the figural mode).

We have employed the term *operative dynamics* here. As noted previously, the "operative dynamics" of the computing machine is the manipulation of symbols in an abstract space and in an abstract time, where the only effective requirement is that the computation be achieved and that only, the concrete dynamics of the computing device underlying the symbol manipulation, whether abacus, PC or super-computer, being irrelevant. The operative dynamics in the preceding list is far more concrete, far more critical.

The previous set is surely not the entire list, and we recognize the difficulty involved. How for example are we going to replicate specification of the scale of time without fully understanding the entire biochemical basis of this scaling, something which is barely yet a subject? How will the COST trajectory be replicated in a robot when (a) Piaget is scarcely studied, and (b) there is clearly a very complex biochemical architecture and brain organization underlying this unfolding trajectory? With the vision of the oscillating brain under the control of cardinal adsorbents, we have just barely broached a vision of something that can support the brain when taken as a reconstructive wave passing through the holographic field, and explicating this conception as a model in physics is a necessary step. Of course, the observant here will note that the requirements scarcely begin to scratch the surface on what is needed to support language and voluntary action.

It is doubtful that on giving these requirements to ChatGPT, it will return a nice set of instructions like that given for the mousetrap. However, perhaps we should take what we would hope is the eventual solution to this as an index of just what really is the power of the human mind.

SPACE AND TIME REVISITED

Throughout this work, we have insisted that the problem of mind, consciousness, and the role of the brain therein cannot be treated within the classic framework of space and time. Recall Gibson's admonition: "The idea that "space" is perceived, and "time" is remembered lurks at the back of our thinking. *But these abstractions borrowed from physics are not appropriate to psychology*" ([Gibson1966], p. 276). AI and cognitive science might feel themselves immune from this admonition; we have attempted to show that they are not. However, there is a question lurking that has not been addressed here, and our sense is that it must be addressed, both to help support the framework that has been presented here, and also to help loosen the grip that the current framework of physical explanation (which is purely mathematical) has on AI, giving the appearance of great support for AI's approach. The question is this: does not the theory of special relativity and its concept of space-time directly conflict with Bergson (and Gibson too)? In the theory of special relativity (henceforth, SR), it is well-known that the very concept of "simultaneity" is refuted. Yet the theory of mind presented here relies on the concept of an "articulated simultaneity" to support the very notion of the symbolic. Bergson's concept of the dynamically, indivisibly transforming Whole, where the motions of objects are "changes or transference of state" inherently implies simultaneity. How can science go forwards within Bergson's conception of time when physics so strongly disagrees?

The rejection of simultaneity is installed, in concrete, in the 4-D "space-time block" of relativity. Two observers in relative motion (Figure 9.1) are held to have different planes of simultaneity, each with a plane upon which all events are in one observer's "present," but some events of observer X's "present" are in observer Y's past, and so on. These different planes imply there is no *universally defined* plane of universal *becoming* on which all points are in the present for all observers in the universe. In other words, this feature of the space-time block has been installed as an *ontological* feature of time. *Becoming*—*creative* becoming—as opposed to the frozen space-time "block," is an intrinsic feature of Bergson's time—yet another contradiction with physics.

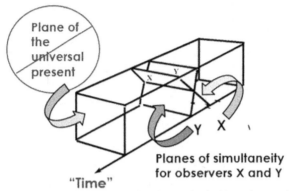

FIGURE 9.1. The 4-D space-time "block" of Special Relativity. Observers in relative motion (X and Y) have different planes of simultaneity (a plane on which all events in the universe are in the "present" for one observer). A common present "plane" of universal "becoming"—a common plane for the entire universe where all next events/points are on a single "future" plane for all observers—seems impossible.

This was a question in the author's mind when a graduate student at the University of Minnesota, working with Robert E. Shaw. Shaw, when already a professor at Minnesota, had been drawn to study with Gibson at Cornell for a year and ultimately founded the journal, *Ecological Psychology*. This question resulted in a paper by the author with the aim of reconciling Bergson with SR, maintaining SR and yet saving at least a form of "becoming." It keyed off a chapter by Milic Capek [Capek1966] in which he examines the Minkowski space-time interval, "*I*", where $I = s - c^2(t_2 - t_1)^2$, s being the 3-D spatial distance between two events at t_1 and t_2. He argued that this invariant quantity defined a tri-partition of event classes which in turn preserves a form of universal becoming (i.e., a definite, invariant causal order or chain issuing from any given point/event in space-time), and as well, defines the relativization of all events *outside* the light cone proceeding from this point as defined by *I*, events that commonsense would otherwise say should be simultaneous. Shaw (who liked this resolution) gave this paper to Gibson who was visiting Minnesota at the time for a conference.

Gibson, in his scribble on the front of the paper, essentially said, "Nice effort, big subject, but maybe look at other things." He gave no reason for this implicitly explicit rejection of the paper's "resolution" at that point. His reason came out in his conference talk the next day, surely in a statement coming "out of the blue" to everyone there, that is, to everyone who had not written that paper, and was very likely unremarked by anyone (well, except by the author). This is Gibson's statement:

> Physicists mislead us when they say there is no simultaneity. When the camera pans to the heroine tied to the rails and then to the hero rushing to the rescue on his horse—these events are simultaneous. (Gibson, 1975, University of Minnesota)

What did this mean? Obviously, Gibson is rejecting the "relativity of simultaneity." The question is: is Gibson right? The short answer is yes. It is likely not known to the readers that Bergson had a public debate with Einstein in 1922 (Cf. [Gunter1969]), a debate that resulted in Bergson's book on the subject, *Duration and Simultaneity*, 1922 (henceforth, D&S [Bergson1922]). When Gibson made his, "Physicists mislead us ..." comment, it was clear that he was siding with Bergson. The debate had centered on the "twin paradox," proposed by Langevin in 1911, now at the core of physics current, ubiquitous, standard interpretation of SR which has installed the relativity of simultaneity as real—as ontological, as a physical fact of the universe, thus unavoidably installing the space-time "block" as a reality as well.

In D&S, Bergson broke down the mathematics of the Lorentz equations in detail along with the system in which Einstein had embedded them. Langevin, Bergson argued, with his rocket-riding twin aging less (actually, physically) than his earth-bound brother, had destroyed the logical consistency and structure of SR. In the reciprocal system in which Einstein had embedded the Lorentz equations (reciprocity: either observer is equally able to claim himself at rest, the other is in motion), all effects are, and can only be, *measurement* effects (cf. A.P. French, 1968, p. 114 [French1968]). What is a measurement effect? I measure my toaster with two rulers; both are of the same length and the same length as the toaster, but one ruler is marked as 6" long, the other 9" long. The toaster "expands" from 6" to 9" depending on the ruler, or contracts (if going from a 9″ ruler to a 6″). Obviously, this is not a real, not an ontological expansion/contraction; it is purely a measurement effect—an effect of the rulers. In SR, the "rulers" are light rays and clocks (synchronized or unsynchronized clocks depending, respectively, on whether the observer is considered at rest or in motion). The Michelson-Morley experiment was "explained" precisely on this basis: while Lorentz had proposed an actual, *physical* contraction of the apparatus arm which lay in parallel with the ether flow, this seemed *ad hoc*; SR was eventually preferred as it indicated there is no actual contraction—it is a measurement effect.

Langevin, with his hypothetical earth-bound twin, now with his cane, long beard and grey hair, very physically aging more rapidly than his rocket-riding, youngish brother, had (a) voided the reciprocity of the two systems (which implied that *either* twin could be construed as aging more), and (b) declared what could only be a measurement effect to be now an *ontological* effect—a real, physical effect, beard, grey hair, and all. He had destroyed SR's logical structure. And note, one cannot declare "time-changes" as ontological and leave length changes as measurement effects (to preserve the Michelson-Morley explanation). In Einstein's system, as Bergson noted, time-units expand exactly in proportion as length-units contract; *the equations are compensatory*. The Lorentz equation for space/length changes and the Lorentz equation for time changes must be of the same order, that is, they must be describing measurement effects.

The "Confirmations"

Thus, and this is what physicists did not want to accept, any use of SR to explain actual "time"-changes (better, rather than "time" changing, the retardation of processes) —the very real slowing of clocks in jets (Hafele-Keating), the longer lifespans of muons with increasing velocity—is invalid. Physics needs a *new theory*, a *different theory* to explain these effects. SR cannot be so used. Yes, the so-called "confirmations" of SR are invalid—a misuse of SR.

Even Bergson's admirers have universally sided with Einstein/Langevin on this subject of the interpretation of SR, ceding Einstein (who sided with Langevin) the argument (i.e., that Langevin's is the correct interpretation). The rock on which even the admirers have floundered is this so-called set of experimental "confirmations" of

SR, or as Canales states things in her exhaustive history of the debate (*The Philosopher and the Physicist* [Canales2015]), the "unambiguous proof," the "stunning confirmation" (p. 31), and Canales, along with everyone else, makes no attempt to note that Bergson was already arguing in 1924 in the context of (and complete awareness of) the earliest of these effects to emerge and considered to be explained by SR, namely, the muons (or mesons in older terminology) with longer lifespans as a function of increasing velocity, even though these are unquestionably valid effects and observations thereof.[2] To him, these very real velocity effects, *in the context of SR's original logical structure*, have no possibility of being considered and used as "proofs" of, or explained by, SR, or as supports of Langevin's interpretation.

We need to repeat the preceding once more, for it is difficult for this point to penetrate the mental defenses of the defenders of the current interpretation of SR: We are *not* denying the validity or reality of these observed effects due to velocity—muon life span increases, slowed clocks on Hafele-Keating's jets. We are not "waving these facts away." We *are* saying, as did Bergson, that SR, *given that its logically consistent structure is understood*, simply *cannot be used to explain these effects*. SR can only explain "measurement" effects. This is not that hard a point, though apparently it is, for the claim will still be made that we are "denying the reality of Hafele-Keating" or something such. Again, physics needs a new theory—a *different theory*—to explain these very ontological effects.[3] It's not

---

[2.] See, for example, Bergson's exchange in *Revue Philosophique* circa 1924 with physicist, Andre Metz (Cf. [Gunter1969], pp. 123–135). Metz could not let the logical problem penetrate; to him *only* SR could explain the life span increase effect on the muon via velocity. Bergson gave up on any further exchange.

[3.] It is for this reason, namely, that the reciprocity is intrinsic to there being any invariance within Einstein's system, that the practice of displaying just *one* Minkowski diagram and arguing from it for SR's explaining the existence of these effects is invalid. One diagram fixes on one observer, say Observer 1, as being at rest, and Observer 2 and all other observers in the universe are set in motion with respect to him. This single diagram proves nothing. We need at least a second diagram to show the symmetric case—Observer 2 now at rest, Observer 1 in motion (which of course undoes the "argument" for Observer 2 being younger due to his motion). Both Bergson (D&S) and A. French [French1968] pointed this out. Yet this argument from a single diagram is ubiquitous in the current literature of SR exposition, with nary a thought of its problems (e.g., [Carroll2022], pp. 146, 165–167).

as though there are no examples; Lorentz himself provided physical, electrodynamical arguments, not just abstract mathematics, for the foreshortening of the apparatus arm parallel to the hypothetical ether flow. Perhaps these can be extended to the retardation of processes (i.e., the slowing of clock mechanisms or the muon life span increases due, say, to the retarding of the nuclear processes by which they radiate themselves away into nonexistence). Albeit it was an ether-based explanation, then again, if the foreshortening also happens to be equally real, equally physical, SR is no longer an explanation of Michelson-Morley.

The foreshortening topic brings up another "paradox," newly coined circa the 1980s, the "pole-barn paradox." The "paradox" notion was now being applied to the length contraction as well. In this paradox, we have a telephone pole which, in its resting state, is too long to fit into a certain barn. However, when the pole is launched into motion at a velocity near the speed of light and flies through the barn, there is a period where the pole, due to its length contraction, fits into the barn. But this paradox is used as a parable for illustrating that we should *not* consider these *real* effects. *The reciprocity is immediately evoked*: It is pointed out that the pole can be seen as at rest, the *barn* could be conceived to be in motion, and therefore it is the barn that will contract. Now the pole does not fit. So, the length contractions are not real, or in philosophical terms, *they have no ontological sta*tus, they are measurement effects. This nicely holds the line with the interpretation of the Michelson-Morley experiment.

However, even the aforementioned obvious contradiction (and logical inconsistency given the compensatory structure of SR) between physics treatment of the "shrinkage of the pole" (i.e., as purely a *measurement effect*) and the retardation of Hafele-Keating's clocks or the twins (as an *ontological effect*) can apparently be confusing, can be simply waved off. One will see responses roughly as follows: "Yes, this is certainly a 'paradox' in the sense that the pole-barn contradicts our prerelativistic intuitions (as do many other relativistic effects), however, it is not a contradiction that the pole is shorter than the barn in some reference frames and longer in others." Ignoring the fact that the only reference frame where the pole "fits" is that of a stationary observer (with or in the barn) who assigns a very high velocity to an observer moving/riding with the pole through the barn and thus, given the pole is now short enough to fit, a very real, very ontological contraction, in other words, (a) it is missed that *physics itself* rejects *the real, ontological contraction* of the pole, and (b) also missed is the concept that a real, ontological contraction of the pole would entail (as physics certainly knows) a total destruction of SR's "explanation" of Michelson-Morley, and (c) since the time and space changes/equations are compensatory, thus of the same order, the time-changes are also just *measurement* effects. To now defend the length contraction as real (which the above response implies) is certainly a strange, misguided, and contrary-even-to-physics mindset.[4]

It would seem that physics could perform the analogy of the pole-barn experiment. One could have an apparatus like a "mini-barn," with perhaps, attosecond light-ray "doors" at each end, which could capture the fact, if it occurred, that a high-velocity "pole" indeed would fit (momentarily) inside the apparatus. Such a finding of course would truly upset current explanatory frameworks.

Saving SR via Multiple Reference Systems

Ignoring the fact that SR's logical structure precludes explaining ontological effects, an SR defender will note that there are arguments that save SR (as an explainer of these effects) *without* an appeal (another favorite move) to *accelerations* being involved due to the rocket's starting and stopping, for example, for non-acceleration arguments, Salmon, 1976 [Salmon1976]; Mermin, 1989 [Mermin1989], as though introducing more complications in terms of scenarios with yet more reference systems in play will correct this intrinsic, structural problem. Salmon was a prototype: Observer *A*

---

[4] Carroll [Carroll2022], in his SR discussion, states: "Length contraction is real, but it's not that the object is physically shrinking, we're just in a different reference frame, measuring a different quantity" (p. 164). So, by "real," does he mean ontological? If he did, SR's Michelson-Morley explanation is no more. But he cannot mean this for he says an object is not *"physically* shrinking," so what does this effect have to be? A *measurement* effect. One or the other. But time and space changes in the equations are compensatory. Now he must refute his support (pp. 146–147) for SR explaining the earth twin's aging as being very real, yes, very ontological, for the earth twin's wrinkles and beard are just not going to "go away" when by reciprocity we argue that it is the earth twin who is moving, not the rocket twin.

leaves earth on a rocket, leaving behind the earth-bound Observer B. On the way out he (A) passes Observer C who is traveling in a rocket from the opposite direction, and A and C exchange signals re the readings of their clocks. C continues on and arrives on earth, and lo, in a triumph for the theory, we see that A's clock was indeed slowed down relative to the earth-bound B's. No accelerations are involved, all explained, supposedly, within the original. uniform-velocity framework of SR.

Except for one problem. Only the *omniscient observer* and creator of the argument (in this case, Salmon) happens to believe A is in motion. A takes with him his own reference system, and A thinks he is at rest, in fact rightfully so via the theory, and B is moving away—and no one, to include all the omniscient observers (who have *zero* status in the framework of the SR theory), can say differently. A's clocks are unchanged (and read so as C passes by). This is exactly, by the way, Bergson's critique (in D&S) of Einstein's general relativity thought experiment with the physicist moving out along the radius of a rotating (accelerating) disk toward the rim, his clock retarding as he goes. Why? Why the retardation? The physicist takes his own reference system along with him. He is at rest; the rim moving toward him. His clock is unchanged. Only Einstein, yet another omniscient observer in this case, thinks the physicist is moving.

In general, all these multiple reference system arguments suffer from this malady: they involve a snapshot in time taken by an omniscient observer who "knows" who is moving, who is at rest. This is not the structure of "relativistic" SR (where *no one* can say who is in motion and who is at rest). And as noted, they are irrelevant to a theory whose logical structure precludes explaining these effects as real, as ontological, and they do not have the wherewithal to (a) resolve the distortion needed to keep the clocks, muons, and twins ontological, but (b) treat as simply a measurement effect the length change of the Michelson-Morley apparatus arm.

Obviously, all this presumes clarity on an elementary confusion: If you are invoking *accelerations* as the mechanism for explaining the time-retardations, thus bringing in the general theory, then your "stunning confirmations" have absolutely nothing to do with SR. You are now "confirming" (this too, very arguably, but not to be addressed here) an explanatory apparatus for "time-retardation" in the general theory (a theory which has nothing *relative* about it) that does not exist within SR. To claim the confirmations are relevant to SR is a clear misdirect.[5]

Simultaneous Causal Flows

All this was Bergson's argument in D&S [Robbins2010], [Robbins2014]. He argued that SR, in its logically consistent structure, with all effects being measurement effects and with its intrinsic reciprocity, preserved the flow of time as an invariant to all observers, that is, neither twin aged more or less than the other. And in this, he pointed out that this invariant flow contains multiple simultaneous causal flows, flows that cannot be relativized.

In his "hero rushing to the rescue, the heroine on the rails" (tied down, struggling, wildly kicking her feet, the locomotive roaring, steaming toward her down the track), Gibson was invoking this "simultaneity of flows." Bergson spent some time in D&S discussing their significance, using illustrations like standing near a river flowing by, with geese flying overhead along the river's course and a rowboat or so being rowed down the stream. Here is one of our own illustrations: Imagine an organically growing, blooming rose. Ultimately make it grow very large, ten feet across. The leading edge of two petals on opposite sides strike two points, one on each side of the blossom (ten feet separated) —and do so simultaneously (to a stationary observer watching the rose). An observer, moving by (but who declares/thinks himself at rest) declares the first observer in motion, his clocks out of sync, the two points *not* being struck simultaneously. This is however absurd: the organic growth of the rose (a simultaneous causal flow) cannot be relativized—not without destroying (at minimum, distorting) the pattern of the organic growth.[6]

[5.] Remember too the discussion in Chapter 6 on the *relativity* of acceleration—the rate of change of the rate of *change of position*—and that it cannot be the *absolute* force assigned in the arguments that invoke the acceleration of the rocket as the cause of "time" slowing. In truth, the Lorentz equations must surely, as Bergson (D&S) noted, be generalizable to acceleration or calculus is useless (see [Wang2003] for a derivation), and then the twin travels to the star and back accelerating and decelerating, the generalized t' is computed, the reciprocity still holds, and there is no differential aging.

[6.] See Rakić's proof [Rakić1997] that Minkowski's schema fails to reflect a correct, clearly logically necessary, historical structure or fabric of events, an argument easily extended by implication to a failure to reflect the causal relations implied in simultaneous flows.

It is not difficult to demonstrate that there are causal consequences to the simultaneity of the flows that simply cannot be made to "go away" depending on the relative motions of observers. As an example, we'll change the rose to a growing tree. Two growing branches strike points on opposite sides of the tree simultaneously, each touching an electric switch that sends a current toward a bell. The bell only rings if both currents with their combined strength hit the bell's switch simultaneously. Presume the bell rings. The ring cannot be made to "go away" (i.e., a valid claim made that it did not happen) simply because a physicist, moving by in a rocket, declares a stationary observer watching the tree to be in motion, his clocks out of sync, and the two branch-strikes "non-simultaneous." The simultaneous, organic, causal flow of the tree's growth cannot be relativized. In other words, the relativity of simultaneity is easily disproved. Gibson, the brilliant expositor of ecological perception, is right.

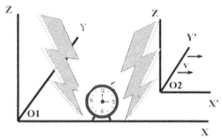

FIGURE 9.2. Two bolts strike simultaneously at 3 p.m. according to stationary observer, O1. O2 is moving away from the bolts with velocity, v, the light from the left bolt lagging behind the right's as it strikes O2's measurement apparatus. To O2, the strikes are not simultaneous.

Even Einstein's iconic, relativized lightning bolts (Figure 9.2) would in fact be part of a large storm front—a vast, boiling, organic system of flows—like the rose. The bolt-strikes can no more be relativized, in truth, than the flows of the rose. The bolt-strikes from within this front are "relative" only via an artificiality, namely, the mathematical characterization of time as a series of "instants," and in this, taking only instantaneous point-events in what is, in reality, a larger flow. The Minkowski schema, with its light cone and its interval, I, represents a causal chain proceeding from an instantaneous point (at the intersection of the past and future light-cones), not a flow. As Bergson noted: "The theoreticians of relativity never note any simultaneity but that of two instants" ([Bergson1922, p. 103). This is but the index of the metaphysic underlying this framework, of which relativity is the logical expression.

A FINAL NOTE ON TIME

The fundamental difference between Einstein and Bergson (and Gibson)—all incapsulated in the argument over SR—is profound. When one sides with Einstein in this subject (and this should be impressed on students of Bergson also), one is inescapably rejecting Bergson's temporal metaphysic—already expressed in *Matter and Memory*—and therefore rejecting his theory of perception and memory. Where it especially comes home to roost, as we have seen, is in the theory of consciousness and the hard problem, especially— as Gibson also saw—that of conscious perception and the origin of the image of the external world.

Bergson counseled the physicists: "A theory of matter is an attempt to find the reality hidden beneath ... customary images which are entirely relative to our needs" ([Bergson1896], p.254). This is equally so and more for the theory of mind. The brain does not dwell in this "customary image," namely, the conceptually imposed abstract space of the classic metaphysic. It could care less about this framework, a framework in fact purely *derivative,* a resultant of the brain's own initial perceptual partition of the environment into "objects" and "motions," the objects being relative to the body's needs, a framework then rarified in the brain's subsequent cognitive developmental trajectory. Rather, the brain and its processes are *intrinsically embedded* in the concrete, indivisible, dynamic transformation described by the temporal metaphysic, and this is the proper framework for starting a theory of mind, consciousness, perception, and ecological intelligence. This has been an underlying theme of this book; it is time to actually consider time.

REFERENCES

[Bergson1896] Bergson, H. *Matter and Memory*. New York: Macmillan, 1896/1912.

[Bergson1922] Bergson, H. *Duration and Simultaneity*. New York: Bobs-Merrill, 1922/1923.

[Canales2015] Canales, J. *The Physicist and the Philosopher.* Princeton, NJ: Princeton University Press, 2015.

[Capek1966] Capek, M. "Time in relativity theory: Arguments for a philosophy of becoming," in J. T. Fraser (Ed.), *The Voices of Time.* New York: Brasiller, 1966.

[Carroll2022] Carroll, S. *The Biggest Ideas in the Universe: Space, Time and Motion.* New York: Penguin Random House, 2022.

[French1968] French, A. P. *Special Relativity.* New York: Norton, 1968.

[Gibson1966] Gibson, J. J. *The Senses Considered as Perceptual Systems.* Boston: Houghton-Mifflin, 1966.

[Gunter1969] Gunter, P. A. Y. *Bergson and the Evolution of Physics.* University of Tennessee Press, 1969.

[Harnad1991] Harnad, S. "Other bodies, other minds: A machine incarnation of an old philosophical problem," *Minds and Machines* (1991): 1, 43–54.

[Hayes1992] Hayes, P., Harnad, S., Perlis, D. & Block, N. "Virtual symposium on virtual mind," *Minds and Machines* (1992): 2(3): 217–238.

[Ingerman1966] Ingerman, P. *A Syntax-Oriented Translator.* New York: Academic Press, 1966.

[Mermin1989] Mermin, D. *Space and Time in Special Relativity.* Long Grove, IL: Waveland Press, 1989.

[Pearl2018] Pearl, J., and D. Mackenzie, *The Book of Why: The New Science of Cause and Effect.* New York: Basic Books, 2018.

[Rakić1997] Rakić, N. "Past, present, future, and special relativity," *British Journal for the Philosophy of Science* (1997): 48, 257–280.

[Robbins2006] Robbins, S. E. "On the possibility of direct memory," in V. W. Fallio (Ed.), *New Developments in Consciousness Research,* New York: Nova Science, 2006.

[Robbins2010] Robbins, S. E. "Special relativity and perception: The singular time of psychology and physics," *Journal of Consciousness Exploration and Research* (2010): *1*, 500–531.

[Robbins2014] Robbins, S. E. *The Mists of Special Relativity: Time, Consciousness, and a Deep Illusion in Physics.* Atlanta: CreateSpace, 2014.

[Salmon1976] Salmon, W. "Clocks and simultaneity in special relativity, or, which twin has the timex?" in P. K. Machamer & R. G. Turnbull (Eds.), *Motion and Time, Space and Matter*. Ohio State University Press, 1976.

[Searle2015] Searle, J. *Seeing Things as They Are: A Theory of Perception*. Oxford, England: Oxford University Press, 2015.

[Shapiro2023] Shapiro, D. "The Birth of Synthetic Intelligence | David Shapiro | Escaped Sapiens #54," May 26, 2023, YouTube video. https://www.youtube.com/watch?v=YfjaspSWI0c&t=14s

[StanfordRoom] https://plato.stanford.edu/entries/chinese-room/, 2023.

[Wang2003] Wang, L. "Space and time of non-inertial systems," *Proceedings of SSGRR*, L'Aquila, Italy, 2003.

[Weiss2002] Weiss, Y., Simoncelli, E., & Adelson, E. "Motion illusions as optimal percepts," *Nature Neuroscience* (2002): 5(6), 598–604.

INDEX